GUIDED INQUIRY EXPLORATIONS INTO ORGANIC AND BIOCHEMISTRY

Revised First Edition

By Julie K. Abrahamson
University of North Dakota

Bassim Hamadeh, CEO and Publisher
Michael Simpson, Vice President of Acquisitions
Jamie Giganti, Managing Editor
Jess Busch, Senior Graphic Designer
Marissa Applegate, Acquisitions Editor
Gem Rabanera, Project Editor
Alexa Lucido, Licensing Coordinator
Mandy Licata, Interior Designer

First published in the United States of America in 2015 by Cognella, Inc.

Printed in the United States of America

ISBN: 978-1-63189-112-0 (pbk) / 978-1-63189-113-7 (br)

www.cognella.com 800-200-3908

CONTENTS

What is Process Oriented Guided Inquiry Learning (POGIL)?

POGIL is a classroom and laboratory technique that seeks to simultaneously teach content and key process skills such as the ability to think analytically and work effectively as part of a collaborative team. POGIL is based on research indicating that (a) teaching by telling does not work for most students, (b) students who are part of an interactive community are more likely to be successful, and (c) knowledge is personal; students enjoy themselves more and develop greater ownership over the material when they are given an opportunity to construct their own understanding. For more information about POGIL, *go to www.pogil.org.*

Group Roles for POGIL Activities

Suggested roles for students in groups include the following. If a group does not have four members, roles may be combined or one role might be omitted.

A **Manager** is designated by the instructor each day, assigns roles in the group, and keeps everyone on task. The manager aims to have all group members participate and understand the concepts discussed. The manager raises a hand if the group has a question.

The **Recorder** records on a Report form the names of each of the group members at the beginning of each day. The recorder keeps track of important observations, insights, group answers, when requested, and other comments.

The **Presenter** presents the work of the group to the class when called upon. Presenters may be asked to go to the board to write out and explain answers or to discuss their answers in comparison to those of other groups.

The **Strategy Analyst** or **Reflector** observes and comments on the group dynamics and behavior regarding the learning process. The analysis may include insights about how well the group worked together to the benefit of all and their learning.

Advice for Success

Be sure you *understand* the answers to the ***Critical Thinking Questions*** and ***Exercises*** in each activity. Listen to the ideas of others in your group to clarify your own. *Ask more questions* until you are confident in your answers. Always read the corresponding sections and work the suggested problems in the text.

1

STRUCTURES AND NAMES OF ALKANES

How are simple organic molecules named and drawn?

Learning Objectives:

- Recognize condensed and structural formulas of alkanes.
- Learn names for alkanes up to 10 carbons.
- Understand alkyl groups and their names.
- Understand constitutional isomers.

Prerequisite Concepts:

- Molecular formulas
- Lewis structures

Information, Part I

Organic molecules are based on carbon structures. **Hydrocarbons**, compounds containing only carbon and hydrogen, include **alkanes**, the simplest of many functional groups. **Functional groups** are the parts of a larger molecule that have characteristic chemical behaviors, and they are used to determine (i) names, (ii) properties, and (iii) reactions for organic molecules.

Table 1.1 Names and Structures of the First Ten Alkanes

Molecular Formula	Name of Alkane	Condensed Structure	Number of Carbons
CH_4	methane	CH_4	1
C_2H_6	ethane	CH_3CH_3	2
C_3H_8	propane	$CH_3CH_2CH_3$	3
C_4H_{10}	butane	$CH_3CH_2CH_2CH_3$	4
C_5H_{12}	pentane	$CH_3(CH_2)_3CH_3$	5
C_6H_{14}	hexane	$CH_3(CH_2)_4CH_3$	6
C_7H_{16}	heptane	$CH_3(CH_2)_5CH_3$	7
C_8H_{18}	octane	$CH_3(CH_2)_6CH_3$	8
C_9H_{20}	nonane	$CH_3(CH_2)_7CH_3$	9
$C_{10}H_{22}$	decane	$CH_3(CH_2)_8CH_3$	10

The structures for the compounds in Table 1.1 are, at times, represented using complete **Lewis structures** in which each atom and its bonds in the molecule are shown. The Lewis structure for ethane is pictured here.

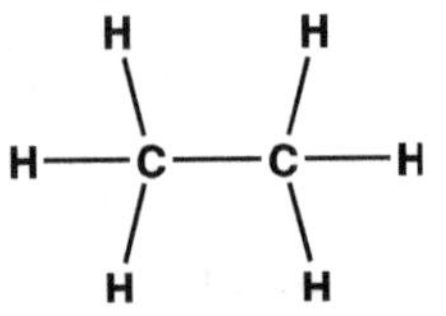

Critical Thinking Questions

1. Look at the molecular formulas for the alkanes. In alkanes with more than two carbons, how could you describe the relationship between the *number of hydrogens* and *carbons*? Compare answers *among your group members.*

2. Draw complete Lewis structures for (a) propane, (b) butane, and (c) heptane.

Information, Part II

When alkanes occur in structures that do not have the carbons connected in a continuous chain, they are described as **branched.** A compound is branched if you cannot trace through all of its carbons without retracing a section or lifting your finger or pencil from the structure. The shorter branching segments of carbons in a compound are **substituents** and are given specific names determined by (a) the number of carbons they contain and (b) in what arrangements they occur.

Table 1.2 Alkyl Groups

Alkyl Group Name	Condensed Structure	Detailed (Lewis) Structure	Number of Carbons
methyl	$-CH_3$		1
ethyl	$-CH_2CH_3$		2
propyl	$-CH_2CH_2CH_3$		3
butyl	$-CH_2CH_2CH_2CH_3$		4

Critical Thinking Question (cont'd.)

3. *As a group*, determine how the alkyl group names and structures in Table 1.2 are related to those of alkanes in Table 1.1 How are they alike or different?

Names of alkanes using the **IUPAC system of nomenclature**, designed for unambiguous naming of compounds, follow these guidelines (with an example on the next page).

IUPAC system of nomenclature

Step 1.
Find the longest *continuous* chain of carbons to name the **parent compound** using an alkane name (see Table 1). (It is helpful to draw a loop around or highlight the longest continuous or "main chain" in the compound and then *circle* any branching substituents.)

Step 2.
Number the carbon atoms in the main chain; give *lowest possible numbers* to substituents (branches or other groups).

Step 3.
Identify and locate the number of the carbon for each branching substituent. If there is more than one of a type of substituent, use prefixes (di-, tri-, etc.) to describe how many are present. Arrange the names of the substituents alphabetically, ignoring prefixes in determining alphabetical order.

Step 4. Write the name as a single word, listing a number before each substituent with the name of the main chain as the last name. When multiple substituents of one type are present, there must be a number to indicate the location of each. Use *commas* between substituent numbers and *hyphens* between numbers and prefixes.

Example 1: Naming a Simple Branched Alkane

1. Make a *loop* around the main chain and *circle* any branching substituents.

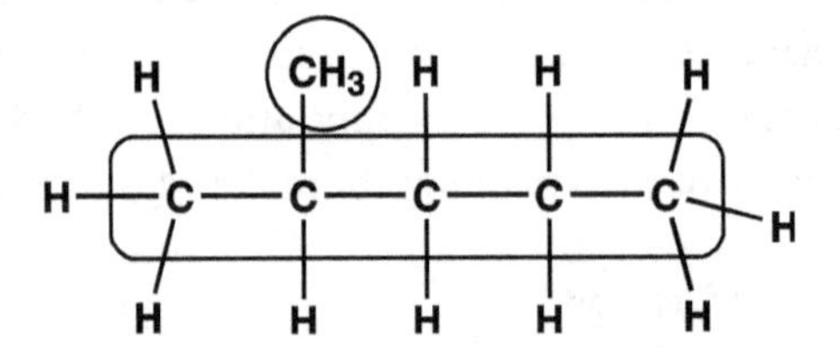

The longest continuous carbon chain has 5 carbons, so the parent compound is ***pentane***.

2. *Number* the carbons in the main chain. Start at the end that gives the branch a lowest number. *Determine* the number of the carbon for the branch (e.g., number 2 in this case).

3. The branch in this compound has only one carbon, so it is called a ***methyl*** group. (*See Table 1.2*) There is only one substituent, so there is no need for any ordering.

4. The IUPAC name is *2-methylpentane* written as one word.

More Examples of Structures and Names

Example 2: 3-ethyl-3-methylhexane
Six carbons in the longest chain, with a methyl and an ethyl substituent, both on carbon 3.

Example 3: 3, 3-dimethylhexane
Six carbons in the longest chain, two methyl group branches on carbon 3.

Critical Thinking Questions (cont'd.)

4. The compounds in Examples 2 and 3 have names ending with *-hexane*. Are these compounds the ***same*** or ***different***? Why or why not?

5. Example 3 has a structure that could be viewed as having a main chain across the middle consisting of five carbons. Why is the five-carbon chain not used in choosing the parent name of the compound?

6. What is the molecular formula for the molecule in Example 1? C____H____

7. What other compound in Table 1.1 has this molecular formula? ______________

8. Are these two compounds the ***same*** or ***different***? (*Circle one.*) Discuss this *as a group*. Come to a conclusion to explain why or why not.

Information, Part III

Compounds that have the same molecular formula but different connections between their atoms are called **constitutional isomers**. These compounds will have *unique* IUPAC names because they have *unique* structures. Isomers often have distinguishable physical or chemical properties.

Critical Thinking Questions (cont'd.)

9. What is the molecular formula for the molecule in example 2? C____H____

10. Which other compound in Table 1.1 has the same molecular formula as Example 2?

11. What is the molecular formula for the molecule in example 3? C____H____

12. Which other compound in Table 1.1 has the same molecular formula as Example 3?

13. Are the compounds in Example 2 and CTQ 10 the ***same*** or ***different***? (*Circle one.*)

14. Are the compounds in Example 3 and CTQ 12 the ***same*** or ***different***? (*Circle one.*)

15. Are the compounds in Examples 2 and 3 the ***same*** or ***different***? (*Circle one.*)

16. *As a group*, determine what is one (or more) simple way to tell whether compounds are the same or different.

17. Which of the compounds in Example 2, Example 3, CTQ 10, and CTQ 12 are *isomers* of each other? Be sure to list any set of isomers. (*Hint:* There may be more than one set of isomers.)

Information, Part IV

There are other substituents that contain *branched* carbon fragments or non-carbon atoms. For convenience in naming, they are given specific names, which will be important to learn. Remember, substituents are *not* part of the longest continuous carbon chain.

Table 1.3 Branched and Halogen Substituent Names

Branched Substituent	Name
CH_3 / $—CH$ / CH_3 *or** CH_3 / $R—CH$ / CH_3	isopropyl
CH_3 / $R—\overset{H_2}{C}—CH$ / CH_3	isobutyl
CH_3 / $R—CH$ / $H_2C—CH_3$	*sec*-butyl
CH_3 / $R—C—CH_3$ / CH_3	*tert*-butyl

Non-Carbon Substituent	Name
–F	fluoro
–Cl	chloro
–Br	bromo
–I	iodo

* On a substituent, **R** indicates where the substituent is attached to the rest (**R** for "rest") of the molecule, i.e., the main chain.

Exercises

1. Complete the table.

Name	Lewis Structure	Condensed Formula	Molecular Formula
2-bromobutane	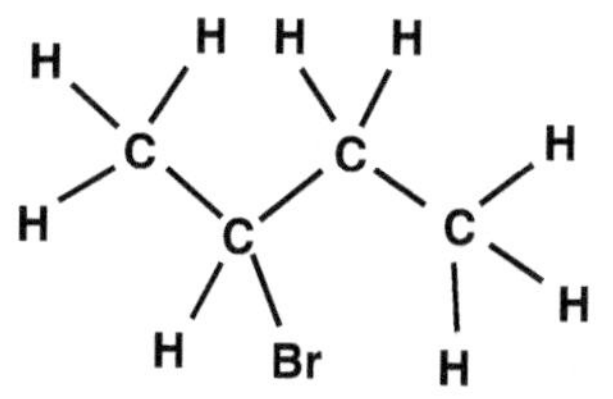		
		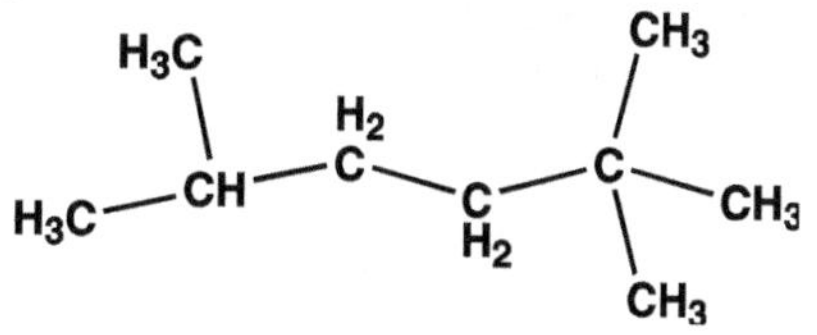	
3-isopropylpentane			
		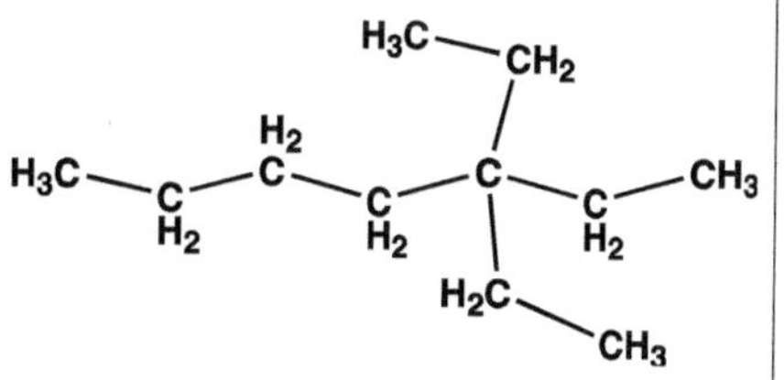	

2. a. *Draw* and *name* as many isomers with the formula C_6H_{14} that you can.
   b. How many isomers are there? __________

3. a. *Draw* and *name* as many isomers of $C_4H_8Cl_2$ as you can find.
   b. How many isomers are there? __________

4. Be sure you *understand* the answers to the ***Critical Thinking Questions*** and ***Exercises*** in this activity. *Ask more questions* until you are confident in your answers.

5. Read the corresponding sections and work the suggested problems in the text.

2

STRUCTURES AND REACTIONS OF ALKENES

How are alkenes different from alkane in structures and reactions?

Learning Objectives:

- Recognize geometric isomers of alkenes.
- Predict products for alkene addition reactions.
- Learn and apply Markovnikov's Rule.
- Explain the differences in reaction types.

Prerequisite Concepts:

- Simple alkene nomenclature
- Constitutional isomers
- Condensed structures
- Functional group identification

Supplies

Molecular model kit

Model 1. Names and Structures of Some Simple Alkenes

Molecular Formula	Name	Condensed Structure	Structural Formula
C_4H_8	*cis*-2-butene	$CH_3CH{=}CHCH_3$	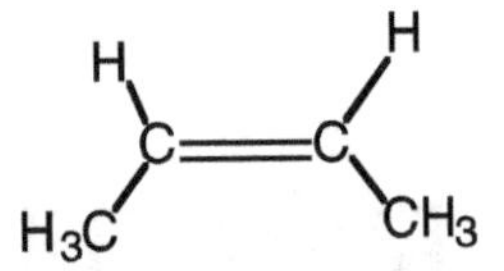
C_4H_8	*trans*-2-butene		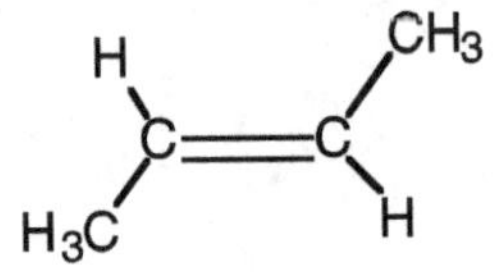
C_5H_{10}	*cis*-2-pentene	$CH_3CH{=}CHCH_2CH_3$	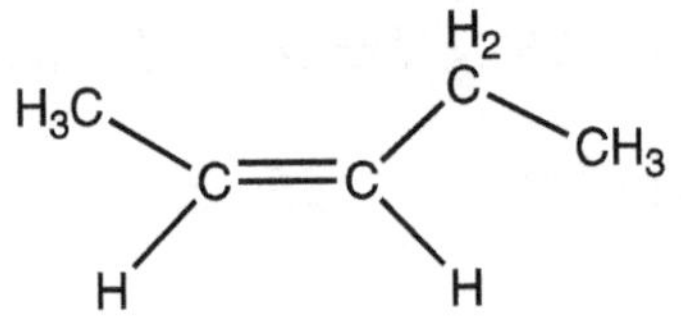
C_5H_{10}	*trans*-2-pentene		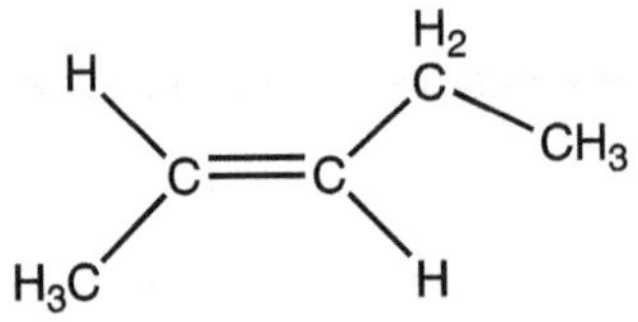
C_6H_{12}	*cis*-3-hexene		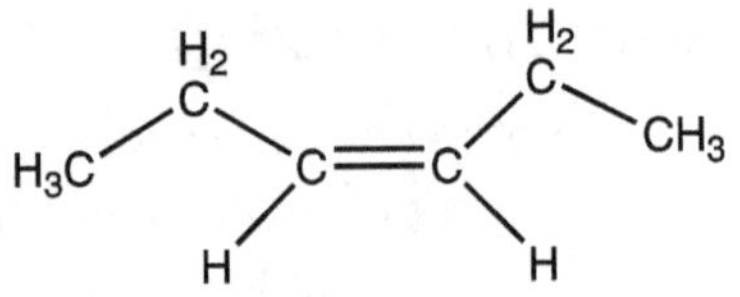
C_6H_{12}	*Trans*-3-hexene		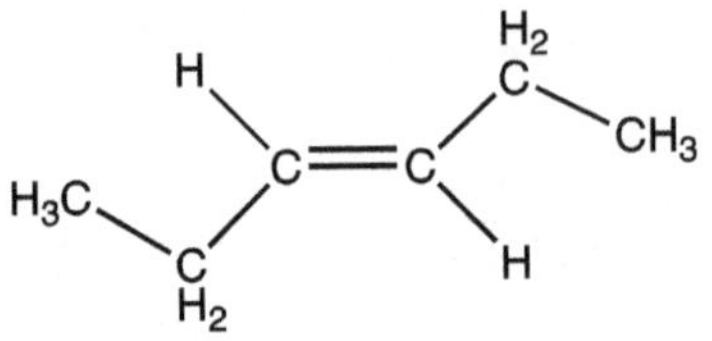

CRITICAL THINKING QUESTIONS

1. Fill in the missing condensed structures in Model 1.

2. Looking at the information in Model 1, how are these compounds related? (*Circle one choice for each pair.*)

a. *cis*-2-butene and *trans*-2-butene	same compound	isomers	unrelated
b. *cis*-2-pentene and *trans*-2-pentene	same compound	isomers	unrelated
c. *cis*-3-hexene and *trans*-3-hexene	same compound	isomers	unrelated

3. a. Draw a loop around the *longest continuous carbon chain* in the *structural* formulas for each compound in Model 1. Compare loops with your group members. Can the loop be drawn more than one way on any of these examples? (*Circle one.*) Yes/No

   b. How could the shape *around the double bond* in the longest carbon chains be described for examples named as *trans*?

   c. *As a group*, decide what the *trans* structures have in common.

4. a. How could the shapes *around the double bond* in the longest carbon chains be described for examples named as *cis*? (*Be creative.*)

   b. *As a group*, decide what the *cis* structures have in common.

5. a. *As a group*, using a **molecular model kit**, make a model of one of the compounds in Model 1.

   b. Using your model, can the compound be converted to the other isomer without breaking a bond? (*Circle one.*) Yes/No

   c. Describe in general terms how the **model** of a *cis*-isomer could be converted to a *trans*-isomer.

Information, Part I

Alkenes that have the same connections between atoms but are different in the geometry around the double bond are called **geometric isomers.** The two forms are called *cis*- and *trans*-isomers.

Model 2. More Alkene Structures

Molecular Formula	Name	Structural Formula	Has *cis-trans* isomers?
C_4H_8	1-butene	H H C=C CH₃ H C H₂	Yes/No
C_5H_{10}	2-methyl-2-butene	H₃C H C=C CH₃ H₃C	Yes/No
C_6H_{12}	3-methyl-2-pentene	H CH₃ C=C CH₃ H₃C C H₂	Yes/No
C_6H_{12}	2-methyl-2-pentene	H₂ C H₃C CH₃ C=C H₃C H	Yes/No
C_7H_{14}	3-methyl-3-hexene	H₂ C H₂ C CH₃ H₃C C=C H₃C H	Yes/No

CRITICAL THINKING QUESTIONS (CONT'D.)

6. a. What is the definition for constitutional isomers? (See Information, Part III, Activity 1, *Structures and Names of Alkanes.*)

   b. Which compound in Model 2 is a constitutional isomer of 2-butene? (*See Model 1.*)

   c. Which compound in Model 2 is a constitutional isomer of 2-pentene?

7. For each compound in Model 2, *draw a loop* around the longest continuous carbon chain in the structural formulas. *As a group,* determine which compound(s) ***could*** have a loop containing the same largest number of carbons drawn more than one way.

Information, Part II

An alkene that has identical groups on ***one end*** (one carbon) of the double bond can have the longest chain drawn two ways without changing its name and ***does not occur as cis- and trans***-isomers. For example, 1-butene has 2 hydrogens on one carbon of the double bond, and for 2-methyl-2-butene or 2-methyl-2-pentene, there are two methyl ($-CH_3$) groups on one end of the double bond. Alkenes that have non-identical groups on ***each*** carbon of the double bond can have the longest chain drawn only one way and ***will occur as cis- and trans***-isomers.

8. Note which of the compounds in Model 2 can be found as *cis*- and *trans*-isomers in the last column. (*Circle one choice for each row.*)

9. *As a group,* describe in general terms how can you tell whether a compound *is able to* exhibit *cis-trans* isomerism.

Model 3. Reactions of Alkenes with Hydrogen or Halogens

$H_2C{=}CH_2$	$+$ H–H $\xrightarrow{\text{Pd, cat.}}$	$H_3C{-}CH_3$
a) Ethene +	hydrogen	→ ethane
$(H_3C)_2C{=}CH{-}CH_2{-}CH_3$	+Br–Br →	$H_3C{-}CH_2{-}CHBr{-}CBr(CH_3)_2$
b) 2-methyl-2-pentene	+ bromine	→ 2, 3-dibromo-2-methylpentane

Critical Thinking Questions (cont'd.)

10. Which functional group is present in the *organic* compounds used as reactants in Model 3?

11. a. Make an **X** through bonds that are broken in the reactants. Use an *arrow* to mark where atoms are added in the reactants forming new bonds.

    b. Describe what happens in these reactions, including which bonds are broken or formed.

12. To what general type of reaction do these reactions belong? (*Circle one.*)

addition elimination rearrangement substitution

13. *As a group,* list the distinguishing characteristics of the reaction type circled in CTQ 12.

Model 4. Reactions of Alkenes with Hydrogen Halides (HX) or Water

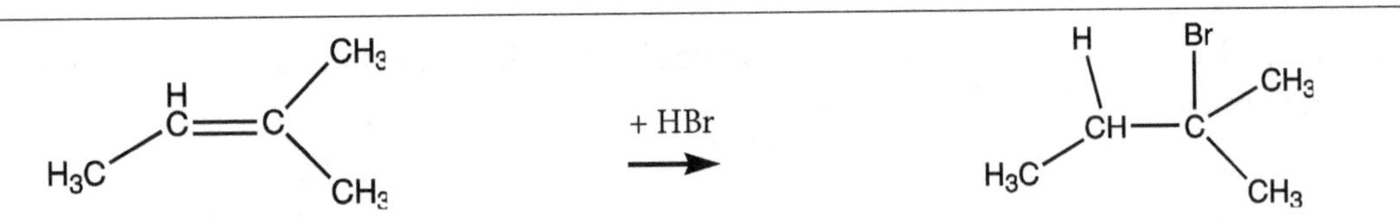

a) 2-methyl-2-butene	+ HBr + hydrogen bromide	→ 2-bromo-2-methylbutane
b) 3-methyl-2-pentene	+H-OH, H_2SO_4,cat. + water	→ 3-methyl-3-pentanol

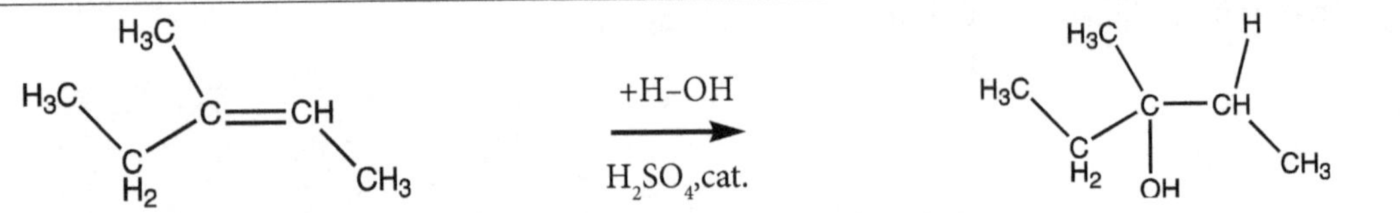

Critical Thinking Questions (cont'd.)

14. To what general type of reaction do the reactions of Model 4 belong? (*Circle one.*)
addition elimination rearrangement substitution

15. What do the reactions in Model 4 have in common?
   a. The H added to the alkene goes to the C of the double bond with ___ H atoms (give the number) *directly* attached. The other C of the double bond has ___ H atoms directly attached.

   b. The H added to the alkene goes to the C of the double bond with ___ H atoms (give the number) *directly* attached. The other C of the double bond has ___ H atoms directly attached.

16. What do the non-organic reactants have in common with those from Model 3? (*Hint:* Think in terms of how many parts the reactants have.)

17. *As a group,* generalize what determines where the H is added in the reactions of Model 4. (*Hint:* Think in terms of the answers to CTQ 15.)

Model 5A. Another Reaction of an Alkene with Hydrogen Halides (HX)

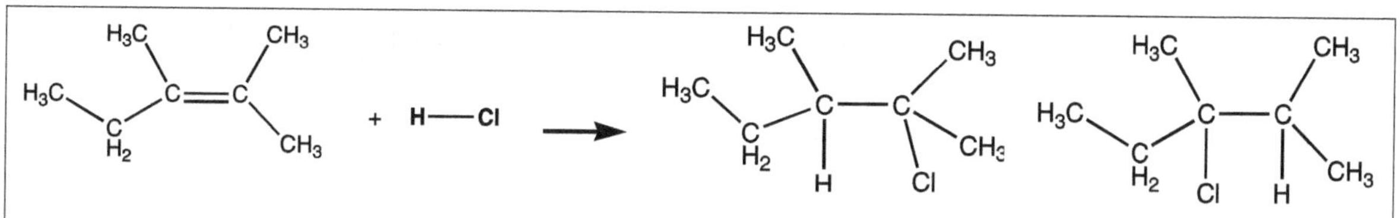

2, 3-dimethyl-2-pentene + HCl → 2-chloro-2, 3- dimethylpentane **Or** 3-chloro-2, 3- dimethylpentane

Model 5B. Another Reaction of an Alkene with Water

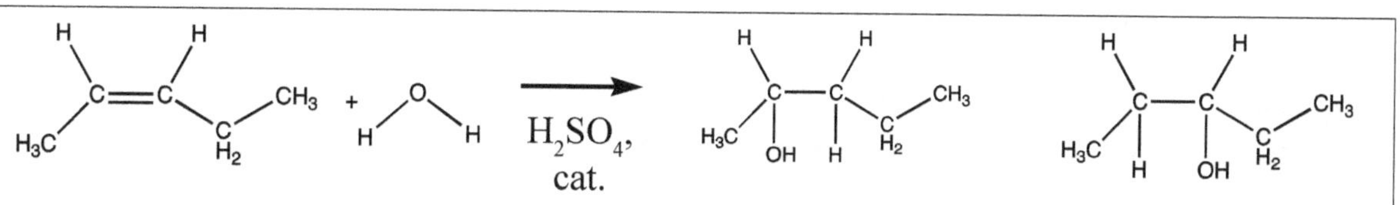

2-pentene + water → (H_2SO_4, cat.) 2-pentanol **Or** 3-pentanol

Critical Thinking Questions (cont'd.)

18. The reactions in Model 5 are similar, yet different, from those in Model 4. How are they similar and different?

    Alike:

    Different:

19. *As a group*, generalize how alkenes add hydrogen plus a halogen or –OH group as shown in Models 4 and 5.

Information, Part III

The reactions of Model 4 show only the major products. The reactions of Model 5 have more than one equally possible product. Those in Model 4 follow **Markovnikov's Rule**, predicting the addition of hydrogen to alkenes from reagents with the formula HX, where X is a halogen or –OH. The rule states that the *major* product results when ***H adds** to the double bond carbon which has the **most H** atoms directly bonded* to it. The reactions of Model 5 have double bond carbons with equal numbers of hydrogens bonded, so the rule does not apply and two possible products result.

Exercises

1. Draw the structural formula for 2-hexene. Does it have *cis-* and *trans-*isomers? (*Circle one.*)
   Yes/No
   Draw and label any *cis-* or *trans-*isomers for 2-hexene.

2. Draw and name one *isomer* of 2-hexene that **does not** have *cis-* and *trans-* isomers. (*Hint*: There is more than one.)

3. Draw and name an isomer of 2-hexene that **does** have *cis-* and *trans-* isomers. (*Hint:* There is more than one.)

4. Write out these reactions. Assume the necessary catalysts are present. How many possible products result for each? *Name* any products you can.

   a. 2-hexene + Cl_2 $\longrightarrow$

   b. 2-hexene + H_2 $\longrightarrow$

   c. 2-hexene + HCl $\longrightarrow$

d. 2-hexene + H_2O ⟶

e. 2-methyl-2-hexene + HI ⟶

f. 2-methyl-2-hexene + H_2O ⟶

5. Be sure you *understand* the answers to the ***Critical Thinking Questions*** and ***Exercises*** in this activity. *Ask more questions* until you are confident in your answers.

6. Read the corresponding sections and work the suggested problems in the text.

3

REACTIONS OF AROMATIC COMPOUNDS

How are aromatic reactions related to those of alkenes?

Learning Objectives:

- Recognize structures of aromatic compounds.
- Predict products of simple aromatic reactions.
- Recognize isomers of aromatic compounds.
- Explain differences in reaction types.

Prerequisite Concepts:

- Functional group identification
- Alkane and alkene reactions
- Constitutional isomers
- Structures of aromatic compounds
- Catalysts

Supplies

Molecular model kit

Model 1A. Reactions of Aromatic Compounds: Halogenation with Bromine

+ Br–Br → (Fe, cat.) Br +H–Br

Benzene + Bromine (Iron catalyst) bromobenzene + HBr

Model 1B. Reactions of Aromatic Compounds: Halogenation with Chlorine

H_2 C CH_3 + Cl_2 → Fe H_2 C CH_3 Cl *Or* Cl H_2 C CH_3 *Or* Cl H_2 C CH_3 + *HCl*

ethylbenzene + chlorine → (Fe, cat) *o*-chloroethyl-benzene *m*-chloroethyl-benzene *p*-chloroethyl-benzene

Critical Thinking Questions

1. What functional group reacts in the *organic* compounds used in Models 1A and 1B?

2. Describe what happens in these reactions including which bonds are broken or formed. (*Hint:* You may find it helpful to use a model kit to represent the reactant and products.)

Information

Reactions of aromatic compounds using other halogens are also *catalyzed* by addition of Fe or $AlCl_3$. Reactions of aromatic compounds with different reagents require catalysts specific to the particular reagent used.

3. How many carbons in the reactant of Model 1A could react equivalently with one Br_2?

4. How many carbons on the aromatic ring of Model 1B can react with one Cl_2? (*Circle them.*)

5. a. *As a group*, determine which carbons of the reactant in Model 1B are equivalent.

   b. Draw and label the reactant below. For each equivalent set, put a symbol by the C to represent the set (e.g.*, Δ, #, etc.). (*Hint:* Use a model kit to represent the reactant and products.)

6. To what general reaction type do the reactions of Model 1 belong? (*Circle one.*)

   addition   elimination   rearrangement   substitution

7. What are distinguishing characteristics of the type of reaction circled in CTQ 6?

8. How are the three possible organic products in Model 1B alike and different?

   Alike: *Different*:

9. How are the three possible organic products in Model 1B related to each other?

10. Why do *three* different products result in the reaction of ethylbenzene with chlorine?

11. Why is the reaction of an *aromatic* compound such as benzene or ethylbenzene with a halogen such as chlorine *different* from that of an *alkene*? (*Hint*: Look at CTQs 1 and 6.)

12. Why does the ethyl group of ethylbenzene remain unchanged in this reaction with chlorine? (*Hint*: Compare reaction conditions in Model 1 to those required for alkane reactions with halogens [e.g., Cl_2].)

Model 2A. Reactions of Aromatic Compounds: Nitration

Toluene (CH_3)	+ HNO_3 Nitric acid	$\xrightarrow{(H_2SO_4,\ cat)}$	o-nitrotoluene (CH_3, NO_2)	+ H_2O

Model 2B. Reactions of Aromatic Compounds: Sulfonation

Toluene (CH_3)	+ H_2SO_4 Sulfuric acid	$\xrightarrow{(SO_3,\ cat.)}$	*m*-toluenesulfonic acid (HO_3S, CH_3)	+ H_2O

Critical Thinking Questions (cont'd.)

13. To what general type of reaction do the reactions of Model 2 belong? (*Circle one.*)

addition   elimination   rearrangement   substitution

14. How many equivalent sets of reactive carbons are present on the ring of the reactant
a) in Model 2A? _____sets   b. in Model 2B? _______ sets

15. *As a group*, decide whether other product isomers to those shown in Model 2 would be possible. Draw the structures for any other possible products.

Exercises

1. Write out the reaction of propylbenzene with Br_2. How many different *mono*bromination products are possible? _______ Draw and name them.

C_6H_5–CH_2–CH_2–CH_3 + Br_2 $\xrightarrow{(Fe,\ cat.)}$

2. Multiple *di*bromination products can result from the reaction in Exercise 1. Draw and name as many as you can.

3. Predict the simplest product(s) of these reactions. (Assume no multiple reactions.)

a) $C_6H_5CH_2CH_3$ $+HNO_3$ $\xrightarrow[H_2SO_4,\ cat]{}$

b) $H_3C\text{-}C_6H_4\text{-}CH_3$ $+ H_2SO_4$ $\xrightarrow[SO_3,\ cat.]{}$

c) $C_6H_5CH_3$ $+I_2$ $\xrightarrow[Fe,\ cat.]{}$

4. a. Which type of reaction is shown in this example? (*Circle one.*)

addition   elimination   rearrangement   substitution

$C_6H_5CH_3 + Br\text{–}Br \xrightarrow[\text{light (catalyst)}]{} C_6H_5CH_2Br + H\text{–}Br$

b. Explain your choice.

c. Why does the reaction occur on the $-CH_3$ group and *not* on the aromatic ring in the reaction shown in Exercise 4?

5. Be sure you *understand* the answers to the ***Critical Thinking Questions*** and ***Exercises*** in this activity. *Ask more questions* until you are confident in your answers.

6. Read the corresponding sections and work the suggested problems in the text.

4

REACTIONS OF ALCOHOLS

How are reactions of alcohols different from those of hydrocarbons?

Learning Objectives:

- Know distinguishing features of addition, elimination, and oxidation reactions.
- Predict major products of alcohol reactions.
- Explain relationship between alkene hydration and alcohol dehydration.

Prerequisite Concepts:

- Functional group identification
- Isomers
- Alcohol nomenclature, classes
- Alkene addition reactions
- Reaction classes

Model 1. Alcohol Dehydration

	Reactants	Catalyst	Major Product(s)	
a)	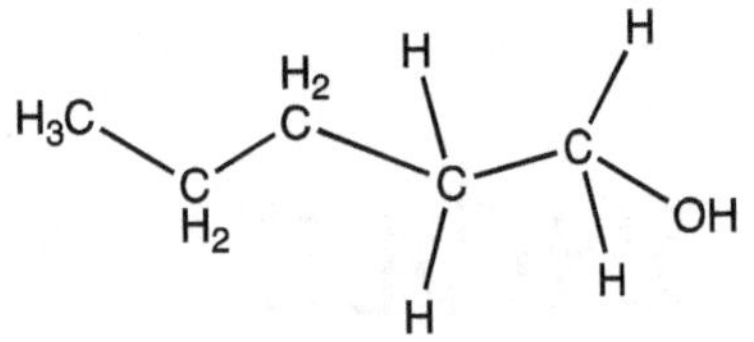 1-pentanol	H_2SO_4 (catalyst)	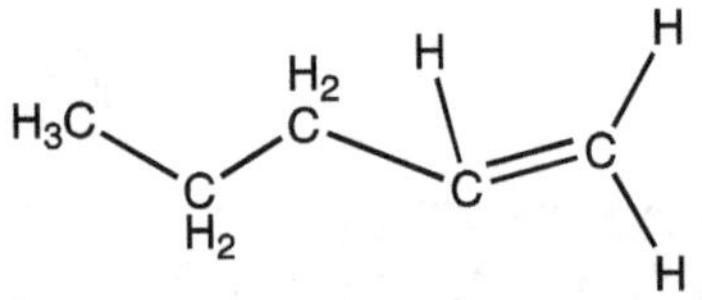 1-pentene	+ H–OH
b)	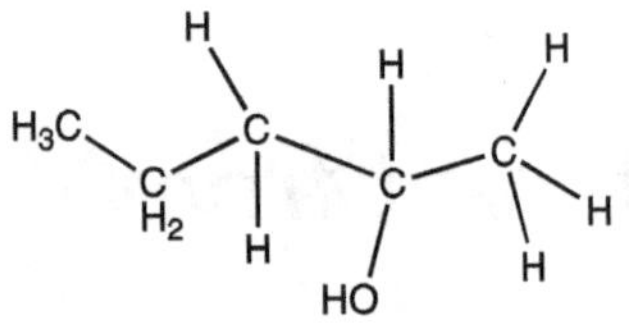 2-pentanol	H_2SO_4 (catalyst)	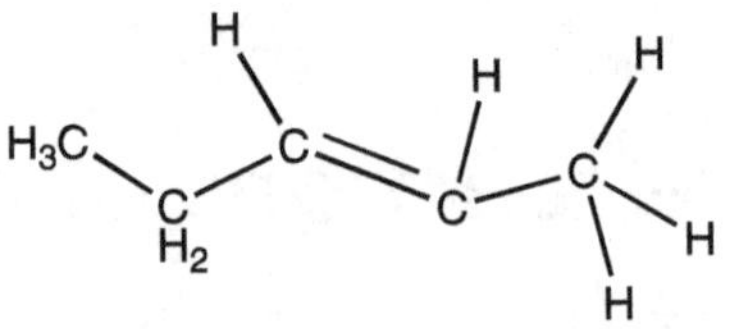 2-pentene	+ H–OH
c)	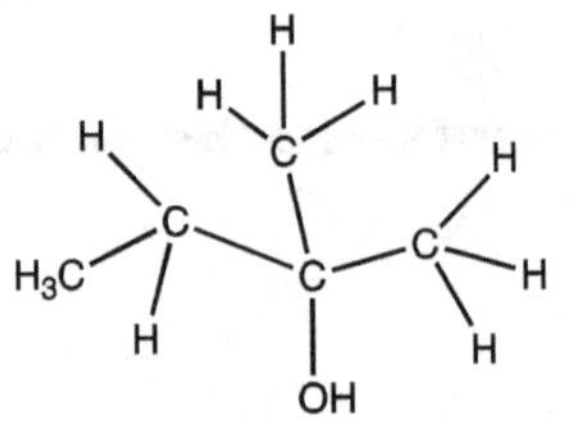 2-methyl-2-butanol	H_2SO_4 (catalyst)	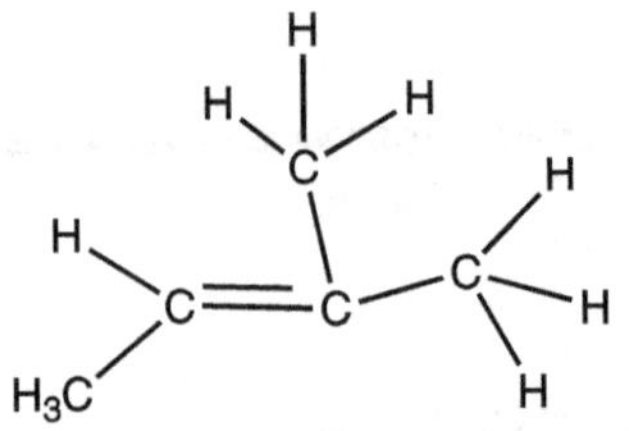 2-methyl-2-butene	+ H–OH

Critical Thinking Questions

1. a. *Circle* the atoms in the reactants from Model 1 that are lost in each process.
   b. Mark an **X** on each bond that is broken.

2. To what general type of reaction do these reactions belong? (*Circle one.*)

   addition   elimination   rearrangement   substitution

3. What are distinguishing characteristics of the type of reaction circled in CTQ 2?

4. *As a group,* determine why these reactions are described as dehydration.

5. How are the reactants in Model 1 related to one another? (*Circle one.*)

   same compound   isomers   unrelated   same functional group only

6. Give the numbers of *hydrogens* directly bonded for each of these:

   a. In Model 1b, on the C to the right of the C with the –OH group: _____

   b. In Model 1b, on the C to the left of the C with the –OH group: _____

   c. In Model 1c, on the C to the right of the C with the –OH group: _____

   d. In Model 1c, on the C to the left of the C with the –OH group: _____

   e. In Model 1c, on the C above the C with the –OH group: _____

7. Complete the statement:

   In the dehydration reactions shown in Models 1b and 1c, the hydrogen is lost from the C *adjacent* to the C with the –OH which has the ***most / fewest*** (*circle one*) hydrogens directly attached.

Information, Part I

Alcohols undergo dehydration reactions in the presence of an acid catalyst such as H_2SO_4 by losing the alcohol –OH group and a hydrogen (H) atom *from a pair of neighboring carbon atoms*. A second bond connects the carbons that lose atoms. When there are neighboring carbons with different numbers of H atoms, as in Models 1b and 1c, the *major* product results when the *H is removed from the carbon with the fewest H atoms.*

Model 2. Alcohol Oxidation

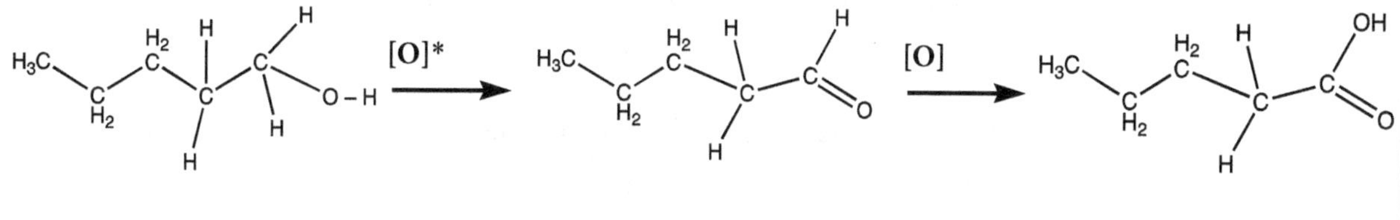

H_3C–CH_2–CH_2–CH_2–O–H $\xrightarrow{[O]^*}$ H_3C–CH_2–CH_2–CH=O $\xrightarrow{[O]}$ H_3C–CH_2–CH_2–C(=O)OH

1-pentanol *[O] = oxidizing agent an aldehyde a carboxylic acid

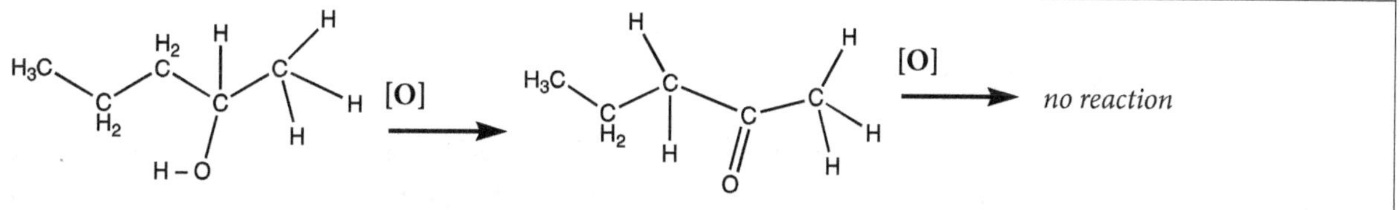

2-pentanol $\xrightarrow{[O]}$ a ketone $\xrightarrow{[O]}$ *no reaction*

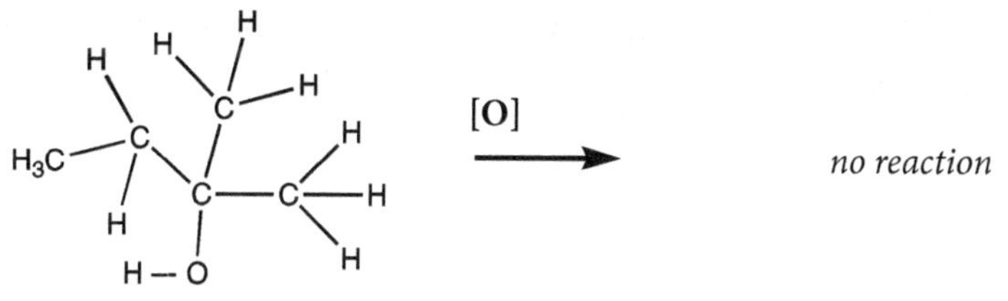

2-methyl-2-butanol $\xrightarrow{[O]}$ *no reaction*

CRITICAL THINKING QUESTIONS (CONT'D.)

8. What type of C is bonded to the alcohol group for each reactant in Model 2? (*Circle one for each.*)

a. 1-pentanol	1°	2°	3°
b. 2-pentanol	1°	2°	3°
c. 2-methyl-2-butanol	1°	2°	3°

9. a. For each *alcohol*, *circle* the atoms lost in oxidation.
   b. For each oxidation product, *draw* a *box* around the carbonyl (C=O) group.

10. *As a group*, confer and describe the positions of the H atoms that are lost relative to the alcohol group for the oxidations of 1-pentanol and 2-pentanol.

11. The 1-pentanol oxidizes to produce an aldehyde. Further oxidation of the aldehyde produces a carboxylic acid. How are these oxidation *steps* different? Describe each step.

12. *As a group*, based on the description of the aldehyde oxidation from CTQ 11, suggest a reason why a ketone does *not* undergo oxidation?

13. The reactions of Model 2 are not classified by the same categories as in CTQ 2. Describe how oxidation occurs for alcohols in terms of which atoms are lost and from where.

14. *As a group*, explain why 2-methyl-2-butanol does *not* oxidize, based on the general description of alcohol oxidation (CTQ 13),

Information, Part II

In organic chemistry, **oxidation** is often described as (a) losing 2 H atoms, (b) gaining an O atom, or (c) increasing the number of bonds between carbon and oxygen. **Alcohols** oxidize in the presence of various oxidizing agents, noted as "[O]," *losing 2 H atoms*, and converting the carbon with the alcohol group to a ***carbonyl carbon*** (i.e., a carbon with a double bond to oxygen). Different types of alcohols generate different carbonyl-containing products.

Concept Check: *Comparing alcohol oxidation to dehydration*

15. a. What reagent is used for oxidation?

    b. Are atoms gained or lost in the process? (*Circle one.*) Gained / Lost

    c. Which atoms are they, and how many?

    d. Describe how different alcohol types are changed in oxidation, including what new functional group results.

16. a. What catalyst is used for dehydration of alcohols?

    b. Are atoms gained or lost in the process? (*Circle one.*) Gained / Lost

    c. Which atoms are they and how many?

    d. Describe how different alcohol types are changed in dehydration, including what new functional group results.

Model 3. Reverse of Alcohol Dehydration: Alkene Hydration

Reactants		Catalyst	Major Product(s)
1-pentene	+ H–OH	H_2SO_4,cat.	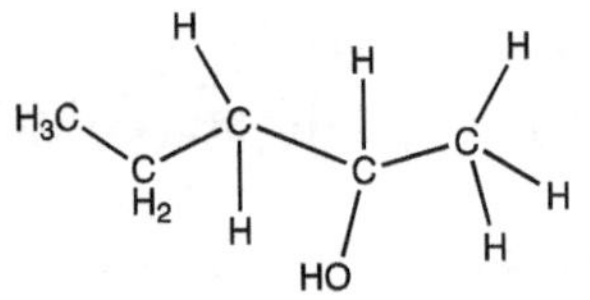 2-pentanol
2-pentene	+ H–OH	H_2SO_4,cat.	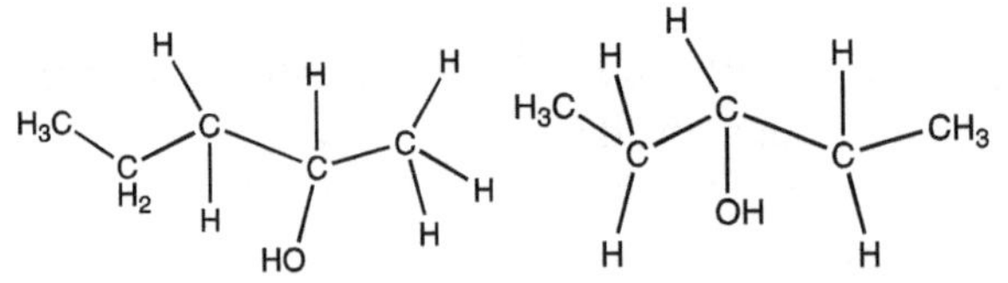 2-pentanol + 3-pentanol
2-methyl-2-butene	+ H–OH	H_2SO_4,cat.	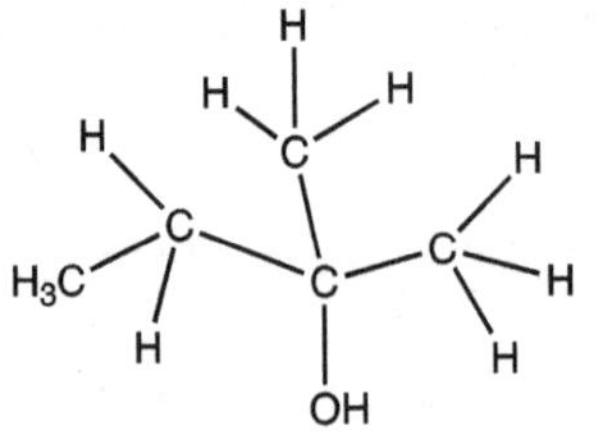 2-methyl-2-butanol

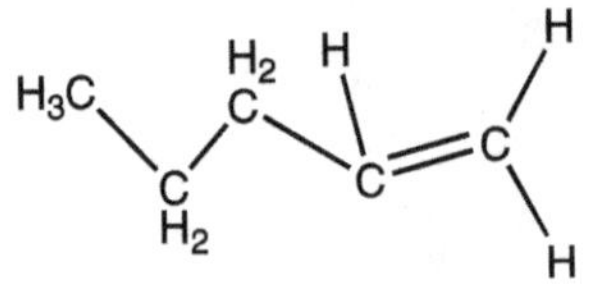

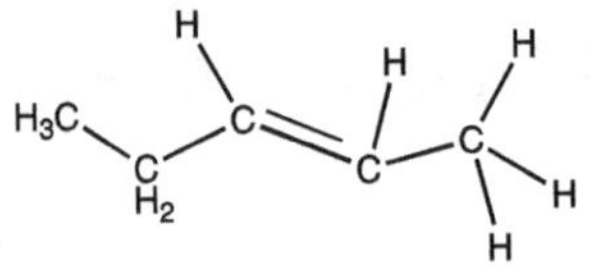

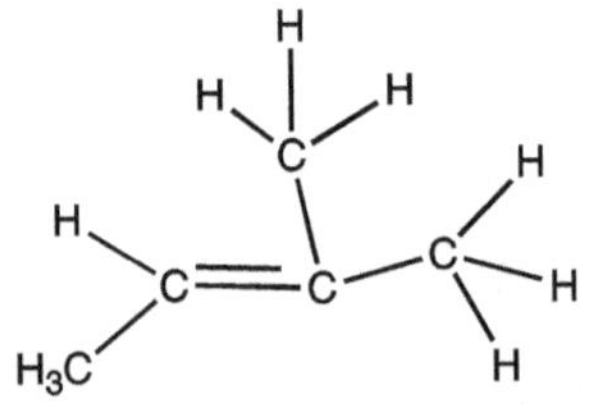

Critical Thinking Questions (cont'd.)

17. a. For each reaction in Model 3, mark an **X** on the bonds that are broken in the reactants.

    b. Mark with *arrows* where new atoms add to the reactants.

    c. *Circle* the new groups added in the products.

18. To what general category do the reactions in Model 3 belong? (*Circle one.*)

    addition   elimination   rearrangement   substitution

19. *As a group,* determine whether any of the reactions of Model 3 are the exact reverse of one shown in Model 1? (*Circle "Yes" or "No."*)

    1-pentene   Yes/No

    2-pentene   Yes/No

    2-methyl-2-butene   Yes/No

20. What is different about any reaction that is *not* the exact reverse of those in Model 1?

Information, Part III: Review of Alkene Reactions

Alkenes undergo **addition reactions** where two atoms are added to the carbons of the double bond, resulting in a product that is no longer ***un***saturated (*Note:* Recall Activity 2, *Structures and Reactions of Alkenes.*) The two added atoms may be *identical*, as for hydrogenation or halogenation, or *different*, as for hydrohalogenation or hydration.

Table 1. Alkene Addition Reactions

Reaction	Reagent	Catalyst	Atoms Added	Product
Hydrogenation	H_2	Pd	H, H	Alkane
Halogenation	X_2	*None*	X, X	Dihaloalkane
Hydration	H_2O	H_2SO_4	H, OH	Alcohol
Hydrohalogenation	HX	*None*	H, X	Alkyl Halide

(*Note:* X is any halogen, e.g., F, Cl, Br, or I.)

For the hydration and hydrohalogenation reactions, more than one possible product can result. The *major product* is determined by Markovnikov's Rule, so that *the H adds to the carbon bound to the most H atoms.* If both carbons of the double bond in the alkene reactant have the same number of hydrogens, both possible products will result.

Exercises

1. Write out the dehydration reaction of 3-pentanol with H_2SO_4. Name any products.

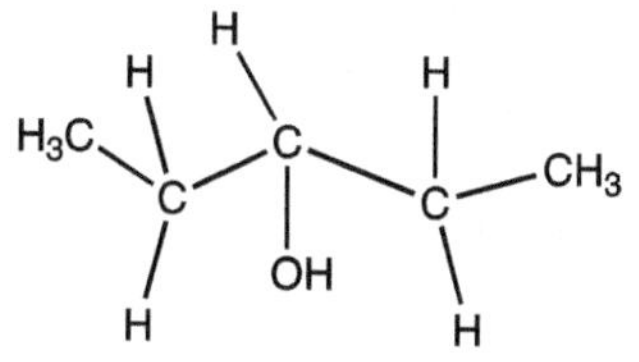

2. a. What type of alcohol is 3-pentanol? (*Circle the answer.*)

   1° 2° 3°

   b. Write out the oxidation of 3-pentanol.

   c. Name the functional group in the product(s).

3. a. Write out the dehydration reaction of cyclopentanol.

   b. Name any products.

4. a. What type of alcohol is cyclopentanol? (*Circle the answer.*)

   1° 2° 3°

   b. Write out the oxidation reaction of cyclopentanol.

   c. Name the functional group in the product(s).

5. a. What type of alcohol is 2, 3-dimethyl-1-butanol? (*Circle the answer.*)

   1° 2° 3°

   b. Write out the dehydration reaction of 2, 3-dimethyl-1-butanol.

   c. Describe the functional group in the product(s), and name any products you can.

   d. Write out the oxidation reaction of 2, 3-dimethyl-1-butanol.

   e. Describe the functional group in the product(s), and name any products you can.

6. a. What type of alcohol is 2, 2-dimethyl-1-butanol? (*Circle the answer.*)

   1° 2° 3°

   b. Write out the dehydration reaction of 2, 2-dimethyl-1-butanol.

   c. Note the functional group in the product(s), and name any products you can.

   d. Write out the oxidation reaction of 2, 2-dimethyl-1-butanol.

   e. Note the functional group in the product(s), and name any products you can.

7. Write out the reaction of 1-pentene with HCl. Name the product(s).

8. Write out the reaction of 2-pentene with HBr. Name the product(s).

9. Be sure you *understand* the answers to the ***Critical Thinking Questions*** and ***Exercises*** in this activity. *Ask more questions* until you are confident in your answers.

10. Read the corresponding sections and work the suggested problems in the text.

5

PHYSICAL PROPERTIES OF ORGANIC COMPOUNDS

What patterns are found in boiling points and water solubilities?

Learning Objectives:

- Understand how intermolecular forces affect physical properties.
- Predict relative boiling points and solubilities based on functional groups.
- Understand patterns in boiling points for different functional groups.

Prerequisite Concepts:

- Functional group recognition
- Structures and names of simple amines
- Amine types: 1°, 2°, 3°
- Isomers
- Intermolecular forces: H-bonding, dipole-dipole, London forces

Part 1. Alkanes, Alcohols, and Ethers

Table 5.1. Physical Properties of Selected Alkanes, Alcohols, and Ethers*

Name	Structural Formula	MW	BP (°C)	Solubility in H_2O
Ethane	CH_3-CH_3	30	-89	insoluble
Methanol	CH_3-OH	32	65	miscible
Propane	CH_3-CH_2-CH_3	44	-42	soluble
Dimethyl Ether	CH_3-O-CH_3	46	-25	soluble
Ethanol	CH_3-CH_2-OH	46	78	miscible
Butane	CH_3-CH_2-CH_2-CH_3	58	0	insoluble
Ethyl methyl ether	CH_3-O-CH_2-CH_3	60	6	soluble
1-Propanol	CH_3-CH_2-CH_2-OH	60	97	miscible
1, 2 Ethanediol	HO-CH_2-CH_2-OH	62	198	miscible
Pentane	CH_3-CH_2-CH_2-CH_2-CH_3	72	36	sl. soluble
Diethyl ether	CH_3-CH_2-O-CH_2-CH_3	74	34	sl. soluble
Methyl propyl ether	CH_3-O-CH_2-CH_2-CH_3	74	39	sl. soluble
1-Butanol	CH_3-CH_2-CH_2-CH_2-OH	74	118	soluble
1, 3-Propanediol	HO-CH_2-CH_2-CH_2-OH	76	215	miscible
Hexane	CH_3-CH_2-CH_2-CH_2-CH_2-CH_3	86	69	insoluble
Methyl butyl ether	CH_3-O-CH_2-CH_2-CH_2-CH_3	88	70	insoluble
1-Pentanol	CH_3-CH_2-CH_2-CH_2-CH_2-OH	88	138	sl. soluble
1, 4-Butanediol	HO-CH_2-CH_2-CH_2-CH_2-OH	90	230	miscible

**Source:* Data from *CRC Handbook of Chemistry and Physics*, 92nd Ed. 2011, CRC Press Inc., Cleveland, OH; Physical Constants of Organic Compounds. Solubility is on a relative scale: insoluble (least), slightly soluble, soluble, very soluble, miscible (greatest).

CRITICAL THINKING QUESTIONS

1. Fill in the boiling points, compounds, and their functional groups using data from Table 5.1.

MW Range	Highest BP	Compound, Functional Group	Lowest BP	Compound, Functional Group
30–32				
44–46				
58–62				
72–76				
86–90				

2. a. What do compounds with the *highest* boiling points for a particular molecular weight range have in common?

   b. What do compounds with the *lowest* boiling points for a particular molecular weight range have in common?

3. Table 5.1 includes three *pairs* of alcohols with similar molecular weights in the ranges of 60–62, 74–76, and 88–90. Compare the boiling points for these pairs. What generalization can be made for each pair?

Information, Part I: Intermolecular Forces and Boiling Points

All compounds have **London forces** that are weak intermolecular attractions resulting from moving electrons yielding instantaneous but brief partial negative (δ–) and partial positive (δ+) charges. These attractions increase as the numbers of electrons increase. Compounds with *similar molecular weights* have similar numbers of electrons and, therefore, *similar London forces.*

Boiling points are correlated to intermolecular forces because these are the forces that must be overcome by addition of heat before a liquid is converted to a gas. Thus, *the greater the intermolecular forces, the higher the boiling point.*

4. Which compounds from Table 5.1 are insoluble or slightly soluble in water? What do most of these have in common?

5. What do all the miscible compounds, i.e., those with the greatest water solubility, have in common?

6. Compare the water solubilities of 1-pentanol to 1, 4-butanediol using the data in Table 5.1. One is miscible, and one is soluble in water. What could account for the difference in the solubilities?

Information, Part II: Intermolecular Forces and Water Solubility

Compounds will be soluble in water when their intermolecular forces are similar to those present in water, i.e., they have dipole-dipole interactions, H-bonding, or both. The more similar the forces are to those for water, the greater the water solubility.

CRITICAL THINKING QUESTION (CONT'D.)

7. Circle the types of intermolecular forces present in compounds with these functional groups.

a. Alkanes	London forces	dipole-dipole interactions	H-bonding
b. Ethers	London forces	dipole-dipole interactions	H-bonding
c. Alcohols	London forces	dipole-dipole interactions	H-bonding
d. Amines	London forces	dipole-dipole interactions	H-bonding

Part 2. Amines

Table 5.2. Physical Properties of Selected Amines*

Amine Type	Name	Structural Formula	MW	BP (°C)	H_2O Solubility
	methylamine	CH_3-NH_2	31	-6.4	v. soluble
	ethylamine	CH_3CH_2-NH_2	45	17	miscible
	propylamine	$CH_3CH_2CH_2$-NH_2	59	47	miscible
	butylamine	$CH_3CH_2CH_2CH_2$-NH_2	73	77	miscible
	hexylamine	$CH_3CH_2CH_2CH_2CH_2CH_2$-NH_2	101	132	slightly soluble
	dimethylamine	$(CH_3)_2$-NH	45	7.3	v. soluble
	ethylmethylamine	CH_3CH_2-NH-CH_3	59	34	v. soluble
	diethylamine	$(CH_3CH_2)_2$-NH	73	55	v. soluble
	methylpropylamine	$CH_3CH_2CH_2$-NH-CH_3	73	63	
	ethylbutylamine	$CH_3CH_2CH_2CH_2$-NH-CH_2CH_3	101	105	miscible
	dipropylamine	$(CH_3CH_2CH_2)_2$-NH	101	108	soluble
	trimethylamine	$(CH_3)_3$-N	59	2.8	v. soluble
	triethylamine	$(CH_3CH_2)_3$-N	101	89	Soluble
	butyldimethylamine	$CH_3CH_2CH_2CH_2$-N-$(CH_3)_2$	101	92	miscible

**Source:* Data from *CRC Handbook of Chemistry and Physics*, 92nd Ed. 2011, CRC Press Inc., Cleveland, OH; Physical Constants of Organic Compounds. Solubility is on a relative scale: insoluble (least), slightly soluble, soluble, very soluble, miscible (greatest).

Critical Thinking Questions (cont'd.)

8. Fill in the first column for Table 5.2 to indicate whether each amine listed is 1°, 2°, or 3°.

9. Consider the boiling points of the 1° amines. Describe briefly how the boiling points correlate with the number of carbons or molecular weight.

10. Consider the boiling points of the 2° amines. Describe briefly how the boiling points correlate with the number of carbons or molecular weight.

11. Consider the boiling points of the 3° amines. Describe briefly how the boiling points correlate with the number of carbons or molecular weight.

12. *As a group*, look at the water solubilities of the compounds in Table 5.2. Determine what the maximum number of carbons is for amines that are mostly soluble (i.e., miscible or v. soluble) in water for these compounds.

Model 1. Hydrogen Bonding in Amines

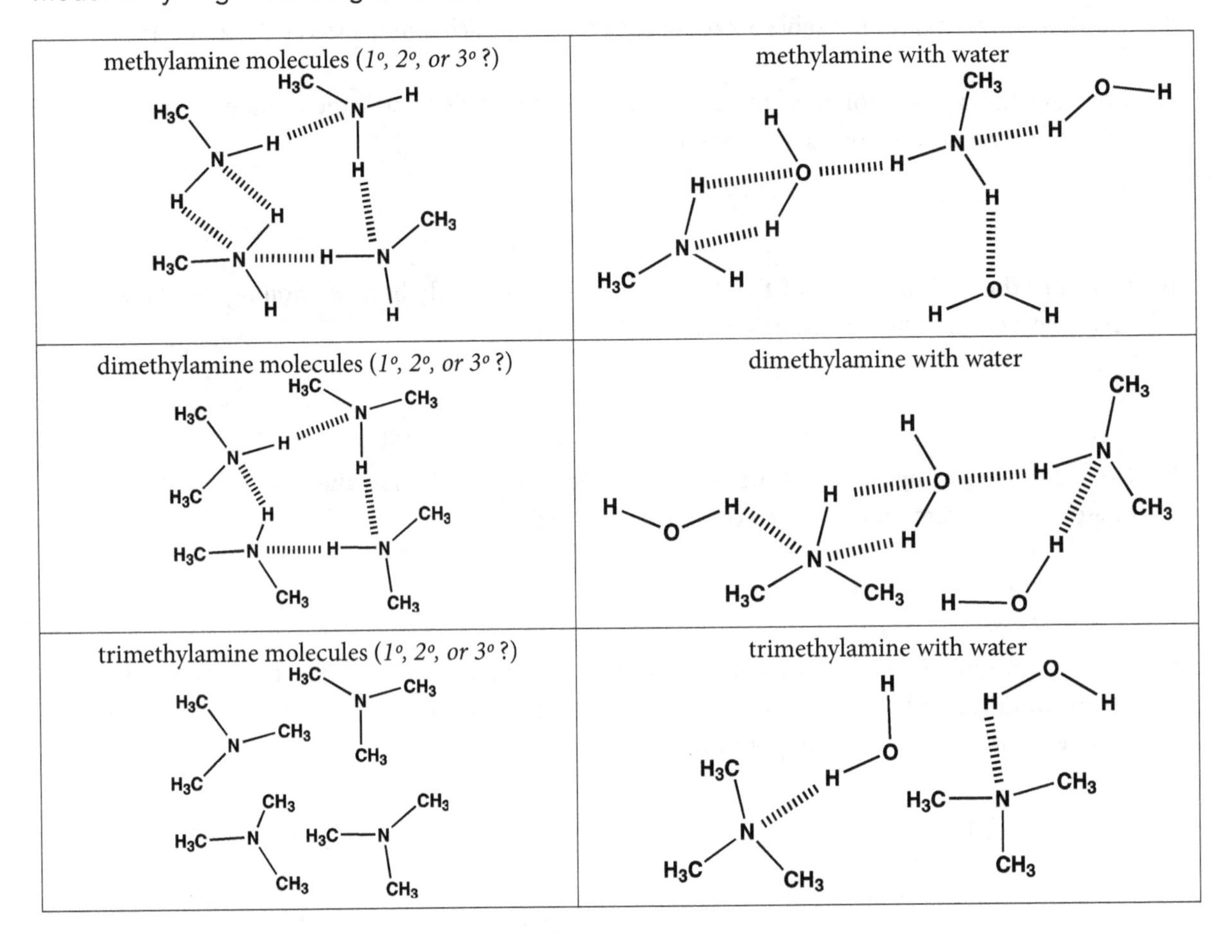

Critical Thinking Questions (cont'd.)

13. a. Add the symbols for partial charges to amine and water molecules in Model 1. There should be "δ–" on each N or O bonded to an H, and "δ+" on each H bonded to an N or O.

    b. *As a group,* determine what *characteristic* is associated with the atoms that are labeled with δ–. Hint: Note their position on the Periodic Table.

14. a. Determine which types of amines (1°, 2°, 3°) are shown in each box of Model 1.
    b. *Circle* the type by each amine name.
    c. How are the H-bonding interactions ***alike or different*** between the three amine types?

    Alike: Different:

15. a. Below is a group of three different amines from Table 5.2 that have the *same molecular weight.* Fill in and compare the boiling points.

MW = 59	propylamine	BP: ________
	ethylmethylamine	BP: ________
	trimethylamine	BP: ________

b. What characteristic(s) of these molecules might account for the differences in their boiling points?

16. a. Next is a second group of three different amines from Table 5.2 that have the *same molecular weight.* Fill in and compare their boiling points.

MW = 101	hexylamine	BP: ________
	butylethylamine	BP: ________
	butyldimethylamine	BP: ________

b. What characteristic(s) of these molecules might account for the differences in their boiling points?

17. How do the London forces compare for the group of compounds compared for CTQ 15?

18. How do the London forces compare for the group of compounds compared for CTQ 16?

19. *As a group*, decide what generalizations can be made for the trends in boiling points of amines based on intermolecular forces. Note both patterns pertaining to amine types and molecular weights (MW). (*Circle the correct word in parentheses.*)

- Boiling points of amines (*increase / decrease*) with increasing MW.
- Boiling points of 1° and 2° amines are (*higher / lower*) than those of 3° amines with similar MWs.
- Boiling points of 1° amines are (*higher / lower*) than 2° amines of similar MWs.

20. Compare the boiling points in Table 5.1 for groups of *alkanes*, *ethers*, and *alcohols*. How can they be generalized? (*Circle the correct word in parentheses.*)

- As the molecular weights of *alkanes* increase, the boiling points (*increase / decrease*).
- As the molecular weights of *ethers* increase, the boiling points (*increase / decrease*).
- As the molecular weights of *alcohols* increase, the boiling points (*increase / decrease*).

21. *As a group*, find the three compounds from Table 5.1 with molecular weights nearest 45. Put an asterisk (*) next to each in the table. Find a 1° or 2° amine from Table 5.2 with the same molecular weight, and mark it with an asterisk. List these compounds in order of increasing boiling points.

    MW Range = ______

Compound Name	____________	< ____________	< ____________	< ____________
BPs	Lowest BP:			Highest BP:

22. Find another group of three compounds from Table 5.1 with molecular weights nearest 73, selecting an alkane, an alcohol, and an ether. Put a triangle (Δ) next to each in the table. Find a 1° or 2° amine from Table 5.2 with the same molecular weight, and mark it with a triangle. List these compounds in order of increasing boiling points.
    MW Range = ______

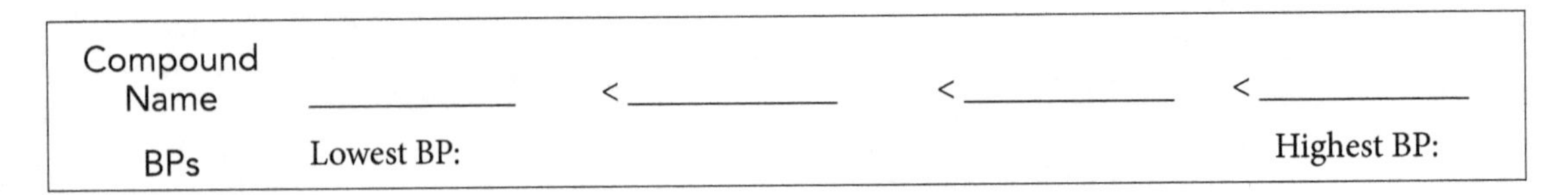

Compound Name	____________	< ____________	< ____________	< ____________
BPs	Lowest BP:			Highest BP:

23. a) *Circle* the *weakest* intermolecular attraction.

    London forces   dipole-dipole interactions   H-bonding

    b) *Circle* the *strongest* intermolecular attraction.

    London forces   dipole-dipole interactions   H-bonding

24. *As a group*, from the patterns noted in CTQ 7, 21, and 22, write a generalized statement about the patterns of boiling points for compounds with *different* functional groups but *similar* molecular weights.

Exercises

1. Draw two molecules of ethylamine, noting where **δ+** and **δ–** occur. Use a dashed line to show where H-bonding attractions would occur between them.

2. Draw molecules of ethylamine and **water**, noting where **δ+** and **δ–** occur. Use dashed lines to show where H-bonding attractions would occur between them.

3. a. To which compounds in Tables 5.1 and 5.2 is the molecular weight of ethylmethylamine most similar? List as many as you find.
   MW: ______

   b. Are the boiling points of these compounds consistent with the generalizations from CTQ 24? Why or why not?

4. a. What term describes how the compounds in CTQ 15 are related to each other?

   b. What term describes how the compounds in CTQ 16 are related to each other?

5. a. Fill in the amine types for the compounds below. Compare the molecular weight and boiling point of 1,2-ethanediamine to those of propylamine from Table 5.2.

Amine Type	Name	Structural Formula	MW	BP (°C)	Solubility in H_2O
	1, 2-ethanediamine	$H_2N\text{-}CH_2CH_2\text{–}NH_2$	60	117	v. soluble
	propylamine	$CH_3CH_2CH_2\text{–}NH_2$			

   b. How do the properties of these compounds compare to the generalizations in CTQ 24?

6. a. Fill in the amine types for the compounds below. Compare the molecular weight and boiling point of *tert*-butylamine to those of butylamine from Table 5.2. Compare also diisopropylamine to dipropylamine from Table 5.2.

Amine Type	Name	Structural Formula	Number of Carbons	MW	BP (°C)	Solubility in H_2O
	tert-butylamine	$(CH_3)_3C{-}NH_2$		73	44.0	miscible
	butylamine	$CH_3CH_2CH_2CH_2{-}NH_2$		73	78	
	diisopropylamine	$((CH_3)_2CH)_2{-}NH$		101	84	slightly soluble
	dipropylamine	$(CH_3CH_2CH_2)_2{-}NH$		101	110	

b. How are the boiling points of these pairs of compounds similar?

7. Table 5.2 does not list information for water solubility of methylpropylamine because none was listed in the *CRC Handbook*. Based on your answer to CTQ 12, which other compounds from Table 5.2 would be expected to have water solubility similar to that of methylpropylamine?

8. Be sure you *understand* the answers to the ***Critical Thinking Questions*** and ***Exercises*** in this activity. *Ask more questions* until you are confident in your answers.

9. Read the corresponding sections and work the suggested problems in the text.

6

PROPERTIES AND REACTIONS OF ALDEHYDES AND KETONES

How does the polarity of the carbonyl group affect these compounds?

Learning Objectives:

- Explain how H-bonding affects properties of aldehydes and ketones.
- Describe the influence of polarity of the carbonyl group on reactions of aldehydes and ketones.
- Predict products of reduction and hemiacetal reactions of aldehydes and ketones.

Prerequisite Concepts:

- Intermolecular forces London forces, dipole-dipole attractions, H-bonding
- Nomenclature of aldehydes and ketones
- Carbonyl group recognition
- Oxidation definitions
- Oxidations of alcohols
- Carbon types: 1°, 2°, 3°

Model 1. Polarity and Attractions in Aldehydes and Ketones

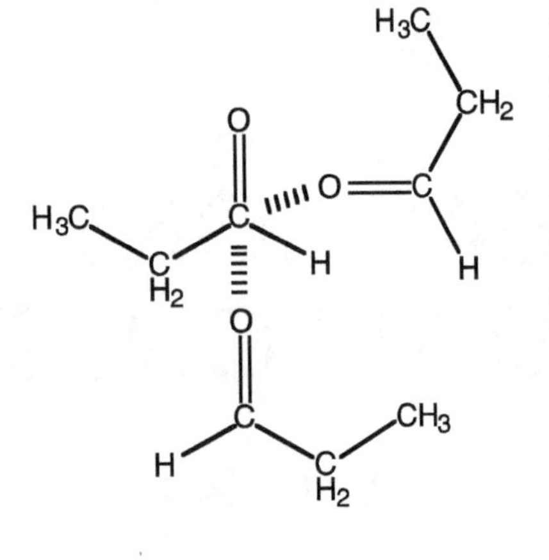 a) attractions between propanal molecules	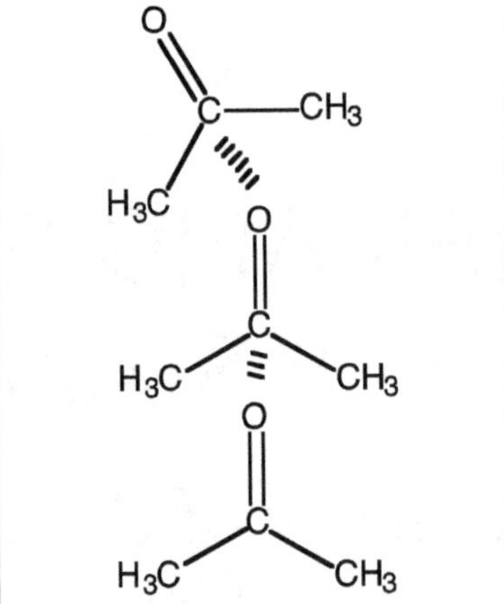 b) attractions between propanone molecules	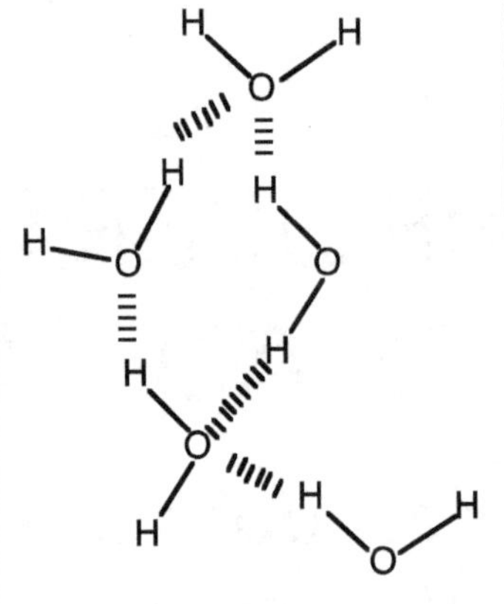 c) attractions between water molecules

Critical Thinking Questions

1. *Mark the dipoles* of the polar bonds in Model 1 by labeling each O atom of the carbonyl groups or water with a (partial) **δ–** symbol. Mark each carbonyl C and H in water with a **δ+** symbol.

2. Which type of attraction(s) occurs between (*Circle those that apply.*)

   a) two aldehyde molecules H-bonding dipole-dipole London forces

   b) two ketone molecules H-bonding dipole-dipole London forces

   c) two water molecules H-bonding dipole-dipole London forces

3. For those places where H-bonding or dipole-dipole attractions occur, between which atoms do the attractions occur? (*Circle those that apply.*)

   a) dipole-dipole O & H O & C O & O H & H

   b) H-bonding O & H O & C O & O H & H

Information, Part I: Review of Intermolecular Forces and Boiling Points

As seen in the previous section (Activity 5, *Physical Properties of Organic Compounds*), ***boiling points*** are determined by the strengths of *intermolecular forces* between molecules. The greater the intermolecular forces, the higher the boiling point. The forces or attractions between molecules include (i) London forces [weakest], (ii) dipole-dipole attractions, and (iii) H-bonding [strongest]. Recall that *H-bonding is an unusually strong attraction between dipoles of O–H or N–H bonds.* (Refer to your text for more details.)

Model 2. Polarity and Attractions Between Aldehydes or Ketones and Water

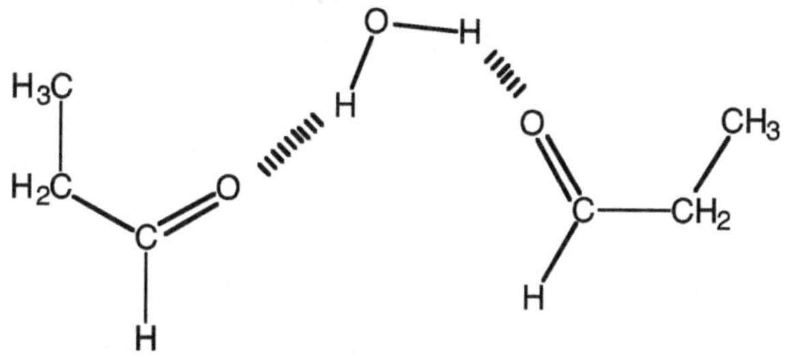 attractions between propanal and water	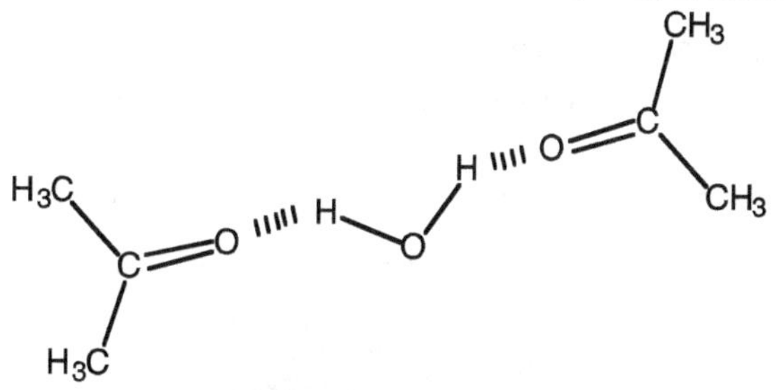 attractions between propanone and water

CRITICAL THINKING QUESTIONS (CONT'D.)

4. *Mark the dipoles* of the polar bonds in Model 2 by labeling all **oxygen atoms** in the water molecules and in the carbonyl groups with a (partial) δ– symbol, and the **hydrogen atoms** of water with a δ+. Mark all the carbonyl carbons with a δ+ symbol.

5. Circle the types of attraction(s) that occur between these molecules.

   a. aldehyde & water   H-bonding   dipole-dipole   London forces

   b. ketone & water   H-bonding   dipole-dipole   London forces

6. For those places in Model 2 where H-bonding or dipole-dipole attractions are found, between which atoms do the attractions occur? (*Circle those that apply.*)

   a. dipole-dipole   O & H   O & C   O & O   H & H

   b. H-bonding   O & H   O & C   O & O   H & H

7. *As a group,* compare the interactions between like molecules (Model 1) to those between propanal or propanone and water (Model 2). Are these compounds expected to be soluble in water? (*Circle "Yes" or "No." Answer "Why or why not."*)

   a. propanone   Yes/No   Why or why not?

   b. propanal   Yes/No   Why or why not?

Information, Part II: Review of Intermolecular Forces and Solubility

The ***solubility*** of compounds in water depends on the presence of sufficient intermolecular attractions between a compound and water. Solubility results when the attractions between two like molecules can be substituted by interactions with a different molecule. The attractions to water molecules will be either H-bonding or dipole-dipole interactions. In some compounds, there is a small attraction to water, which is counteracted by repulsions with water molecules. The negative effects can be a result of non-polar portions of a molecule that repel water and are not attracted to water.

Model 3. Reduction of Aldehydes and Ketones

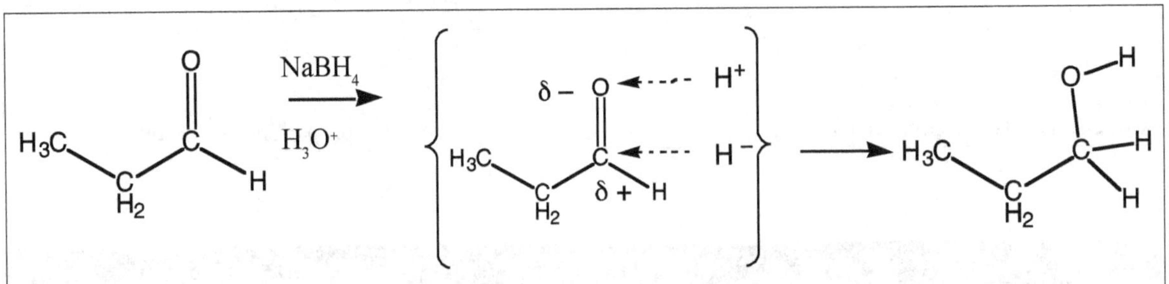

$NaBH_4$, H_3O^+

propanal + sodium borohydride, H_3O^+ *{how the atoms are attracted}* → 1-propanol

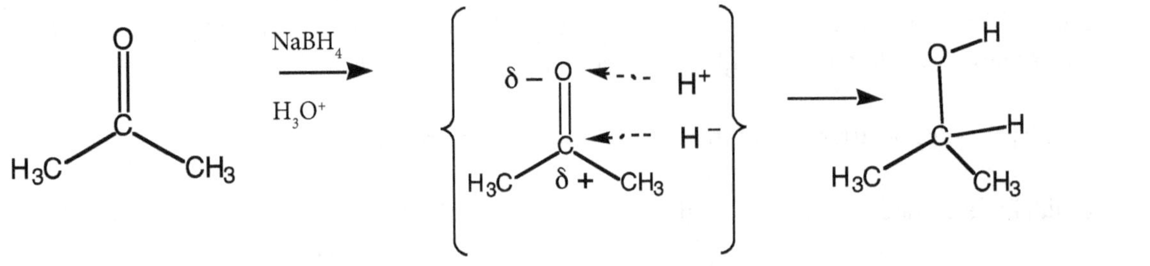

$NaBH_4$, H_3O^+

propanone + sodium borohydride, H_3O^+ *{how the atoms are attracted}* → 2-propanol

CRITICAL THINKING QUESTIONS (CONT'D.)

8. Mark an **X** through any bond that is broken in the propanal and propanone for each reaction in Model 3. *Circle* the atoms that were added in each product.

(Circle the answers for CTQ 9–12.)

9. What type of carbon is the carbonyl C in propanal? 1° 2° 3°

10. What type of alcohol results from the reduction of propanal? 1° 2° 3°

11. What type of carbon is the carbonyl C in propanone? 1° 2° 3°

12. What type of alcohol results from the reduction of propanone? 1° 2° 3°

Information, Part III: Reduction

In organic chemistry, **reduction** is often seen as (a) addition of 2 H atoms, (b) losing an O atom, or (c) decreasing the number of bonds between carbon and oxygen. Reduction is the reverse of oxidation. (Recall the definitions for oxidation.) Aldehydes and ketones are reduced to yield alcohols. The type of alcohol that results depends on the starting carbonyl containing compound. Recall that 1° alcohols oxidize to yield aldehydes, while 2° alcohols oxidize to yield ketones. (*Note:* Refer to Activity 4, *Reactions of Alcohols*, for the definitions for oxidation.)

Aldehydes and ketones can be reduced using the reagent $NaBH_4$ (sodium borohydride) in aqueous acid (H_3O^+). The borohydride provides an $\mathbf{H^-}$ (hydride ion) and the H_3O^+ provides $\mathbf{H^+}$; both are attracted to the ends of the polar carbonyl with the opposite polarities. The double bond of the carbonyl is broken and 2 H atoms add to the molecule, one to each atom of the carbonyl, resulting in the formation of an alcohol. The byproducts $NaBH_3$ and H_2O are generally ignored because, typically, only organic products are of interest.

Concept Check: *Find answers in the Information section above.*

13. Give three definitions of reduction.

    a.

    b.

    c.

14. Reduction is the reverse of ______________________________.

15. Aldehydes and ketones reduce to form ______________________________.

16. What reagent is used for reducing aldehydes and ketones (two parts)?

17. Which bond breaks in the aldehyde or ketone in reduction?

18. Which atoms add where the bond was broken?

Model 4. Alcohol Reaction with Aldehydes and Ketones

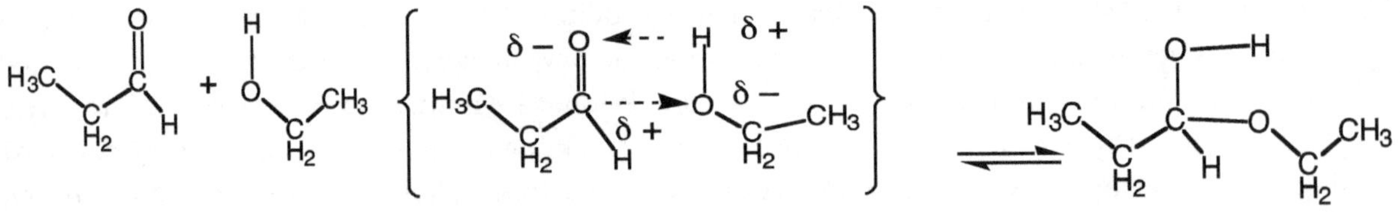

(a) propanal + ethanol *{how the molecules are attracted}* ⇌ a hemiacetal

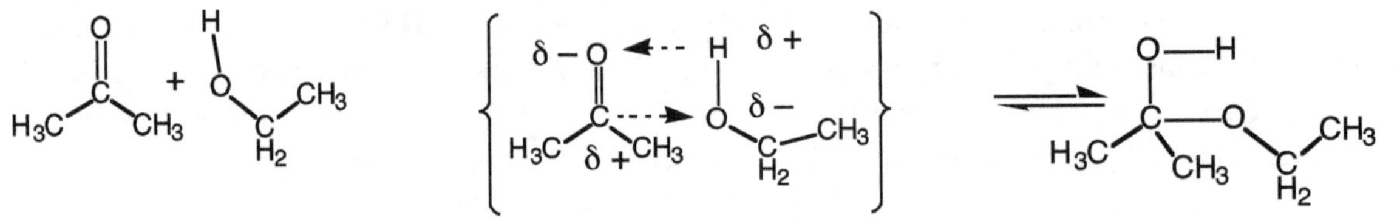

(b) propanone + ethanol *{how the molecules are attracted}* ⇌ a hemiacetal[§]

§ Some sources will distinguish the product of an alcohol reacting with a ketone from that with an aldehyde by using different names. The product using a ketone may be called a *hemiketal.*

CRITICAL THINKING QUESTIONS (CONT'D.)

19. The formation of a hemiacetal is a reversible process. Circle the reaction type.

    addition   elimination   rearrangement   substitution

20. Mark an **X** through the bonds that are broken in the reactants in the reactions of Model 4. *Circle* the new bonds that were formed in making each product. Put a *box* around the carbons in the product that came from the alcohol reactant. Put an *asterisk* (*) by the carbon that was originally in the carbonyl.

21. The asterisk (*) marked carbon is called the *hemiacetal* carbon. *As a group*, list the groups attached to this hemiacetal carbon in the product from the aldehyde (Model 4a).

    ____________ ____________ ____________ ____________

    In this list, *circle* the groups that were originally attached to this (*) same carbon in the aldehyde.

22. List the groups attached to the *hemiacetal*[§] carbon in the product formed from the ketone (Model 4b).

    ____________ ____________ ____________ ____________

    In this list, *circle* the groups that were originally attached to this (*) same carbon in the ketone.

23. Which group(s) is *different* in the hemiacetal formed from the aldehyde compared to that from the ketone? *Discuss your answers as a group.*

Information, Part IV: Hemiacetals

A hemiacetal has both an alcohol (–OH) and an ether (–OR) attached to the same carbon. When the hemiacetal results from reaction with an aldehyde, one hydrogen remains on the hemiacetal carbon. When the hemiacetal results from reaction with a ketone, there is no hydrogen on the hemiacetal carbon.

Exercises

1. Draw two molecules of pentanal showing where attractions between them will occur. Be sure to label the atoms of the polar bonds with **δ+** and **δ–** symbols where appropriate.

2. Draw a representation of the interactions between pentanal and water, showing where attractions between them will occur. Be sure to label the atoms of the polar bonds with **δ+** and **δ–** symbols where appropriate.

3. a. Is pentanal likely to be soluble in water? (*Circle one.*) Yes/No
   Why or why not?

   b. If yes, will it be *more / less* soluble in water than propanal? Explain your answer.

4. Draw two molecules of 2-pentanone showing where attractions between them will occur. Be sure to label the atoms of the polar bonds with **δ+** and **δ–** symbols where appropriate.

5. Draw a representation of the interactions between 2-pentanone and water, showing where attractions between them will occur. Be sure to label the atoms of the polar bonds with **δ+** and **δ–** symbols where appropriate.

6. a) Is 2-pentanone likely to be soluble in water? (*Circle one.*) Yes/No
   Why or why not?

   b) If yes, will it be more / less soluble in water than propanone? Explain your answer.

7. Write out the **reduction** of pentanal using $NaBH_4$ and H_3O^+.

   What type of carbon is in the carbonyl of the reactant? 1° 2° 3°

   What type of alcohol is formed in the product? 1° 2° 3°

8. Write out the **reduction** of 2-pentanone using $NaBH_4$ and H_3O^+.

   What type of carbon is in the carbonyl of the reactant? 1° 2° 3°

   What type of alcohol is formed in the product? 1° 2° 3°

9. Write out the reaction showing the addition of ethanol to butanal. Write in where the partial charges (**δ**+ and **δ**–) occur in the reactants. Mark the hemiacetal C with an asterisk (*).

10. Write out the reaction showing the addition of methanol to 3-pentanone. Write in where the partial charges (**δ**+ and **δ**–) occur in the reactants. Mark the hemiacetal C with an asterisk (*).

11. Be sure you *understand* the answers to the ***Critical Thinking Questions*** and ***Exercises*** in this activity. *Ask more questions* until you are confident in your answers.

12. Read the corresponding sections and work the suggested problems in the text.

7

CONDENSATION REACTIONS OF CARBOXYLIC ACIDS

What similarities are there between reactions of carboxylic acids?

Learning Objectives:

- Describe carbonyl group substitution reactions.
- Predict esters and amides produced from carboxylic acids.
- Predict hydrolysis products of esters and amides.
- Name esters and amides using IUPAC system.

Prerequisite Concepts:

- Functional group recognition
- Substitution reactions
- Catalysts
- Names of carboxylic acids, amines
- Polymers

Model 1. Amide Formation

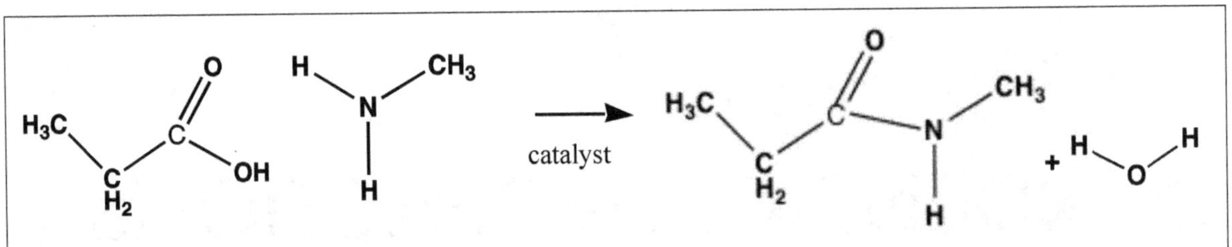

a) propanoic acid	+ methylamine	catalyst →	N-methylpropanamide + water
b) ethanoic acid	+ dimethylamine	catalyst →	N, N-dimethylethanamide + water

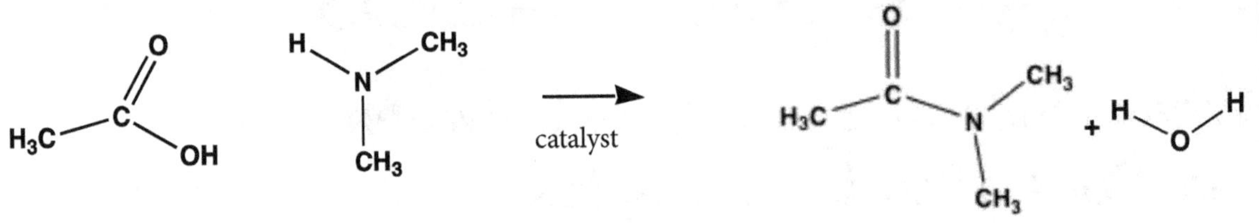

Critical Thinking Questions

1. Draw an **X** through the bonds that are broken in the reactants of Model 1. *Circle* the H and OH in the reactants that are the sources of H_2O in the product. Draw a *box* around the *amide* group in each product.

2. Does the carbonyl group itself react or change in these processes? (*Circle one.*) Yes/No

3. *As a group,* determine how the formation of an *amide* fits the description of a "carbonyl group substitution reaction."

4. a. Which reactant provides the *first part* of the name of the product?

   carboxylic acid   amine

   b. How does its name change in the product?

   c. Which reactant provides the *middle* of the name (the part before "-amide") of the product?

   carboxylic acid   amine

   d. To what new ending does the name of this reactant change in the product? ________

Information, Part I: Amide Catalysis

Heat or catalysts are required for the formation of amides whether in a lab or physiologically. Experimentally, e.g., in a procedure for preparation of nylon, specific catalysts or reactive acid derivatives such as acyl chlorides or acid anhydrides are used in place of carboxylic acids.

Model 2. Ester Formation

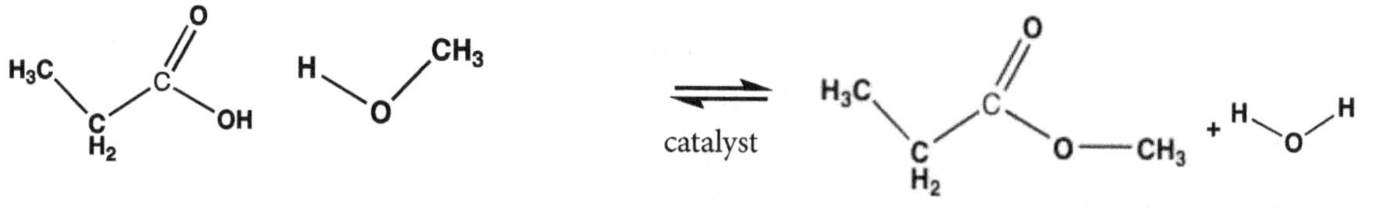

a) propanoic acid	+ methanol	$\rightleftharpoons$ catalyst	methyl propanoate + water
b) ethanoic acid	+ butanol	$\rightleftharpoons$ catalyst	butyl ethanoate + water

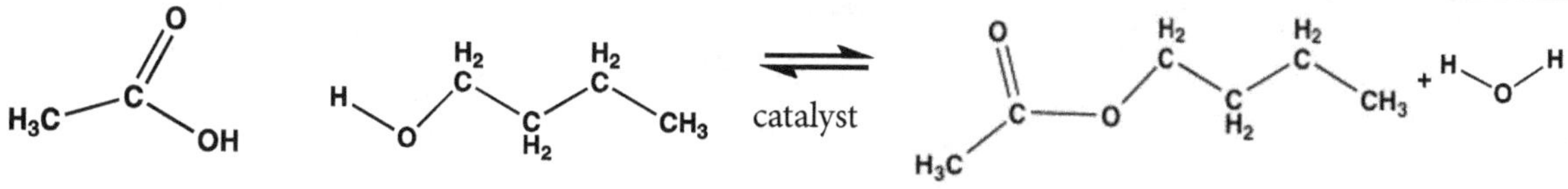

CRITICAL THINKING QUESTIONS (CONT'D.)

5. Draw an **X** through the bonds that are broken in the reactants of Model 2. *Circle* the H and OH in the reactants that are the sources of H_2O in the product. Draw a *box* around the *ester* group in each product.

6. Does the carbonyl group itself react or change in these processes? (*Circle one.*) Yes/No

7. *As a group*, determine how the formation of an *ester* fits the description of a "carbonyl group substitution reaction."

8. a. Which reactant provides the *first* name of the product?

   carboxylic acid alcohol

   b. To what new ending does the name of this reactant change in the product?

   c. Which reactant provides the *last* name of the product?

   carboxylic acid alcohol

   d. To what new ending does the name of this reactant change in the product?

Information, Part II: Ester Catalysis

Acid catalysis is required for the formation of esters in a lab, and enzyme catalysis occurs in biological systems. Experimentally, e.g., in a procedure for preparation of aspirin, reactive acid derivatives such as acyl chlorides or acid anhydrides are often used in place of carboxylic acids.

Model 3. Hydrolysis of Esters

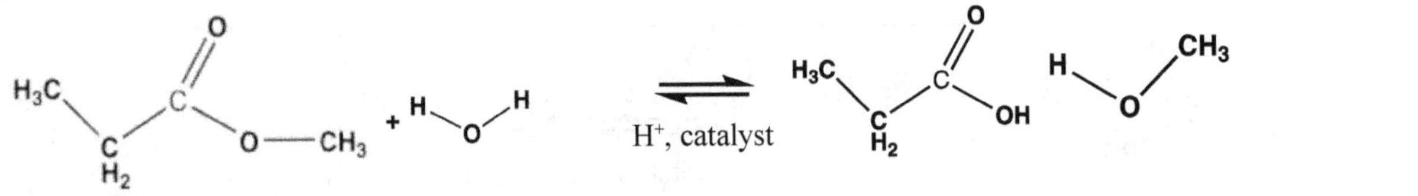

Reactants	Conditions	Products
a) methyl propanoate + water	H^+, catalyst	propanoic acid + methanol
b) butyl ethanoate + water	H^+, catalyst	ethanoic acid + butanol

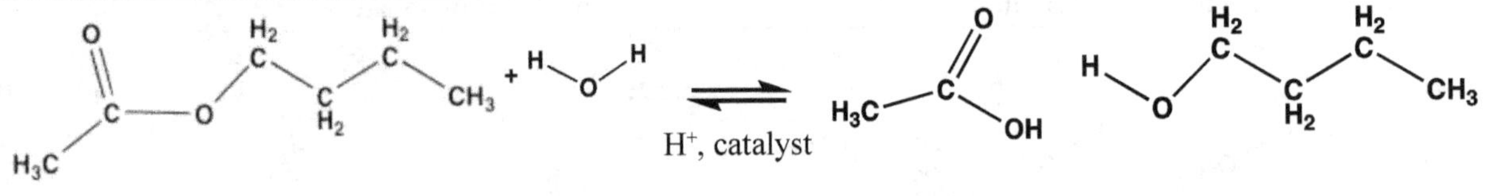

Critical Thinking Questions (cont'd.)

9. Draw an **X** through the bond in each reactant in Model 3 that is *broken*. *Circle* the H and OH groups from water that appear in the products. Which type of bond breaks in the organic reactants?

10. Does the carbonyl group itself react or change in these processes? (*Circle one.*) Yes/No

11. *As a group,* determine how the hydrolysis of an ester fits the description of a "carbonyl group substitution reaction."

Model 4. Hydrolysis of Amides

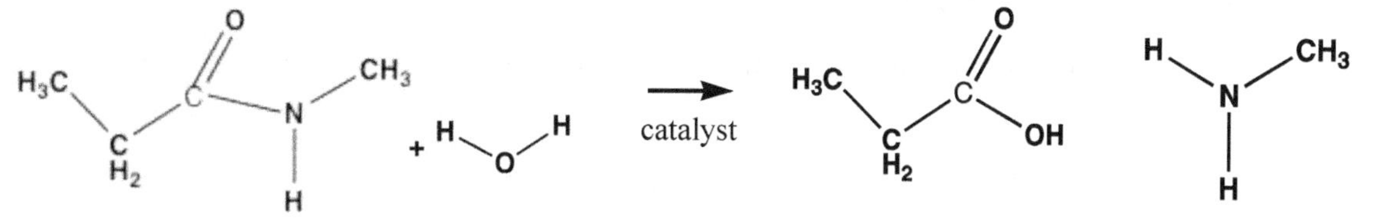

Reactants		Products	
a) N-methylpropanamide + water	$\xrightarrow{\text{catalyst}}$	propanoic acid	+ methylamine
b) N, N-dimethylethanamide + water	$\xrightarrow{\text{catalyst}}$	ethanoic acid	+ dimethylamine

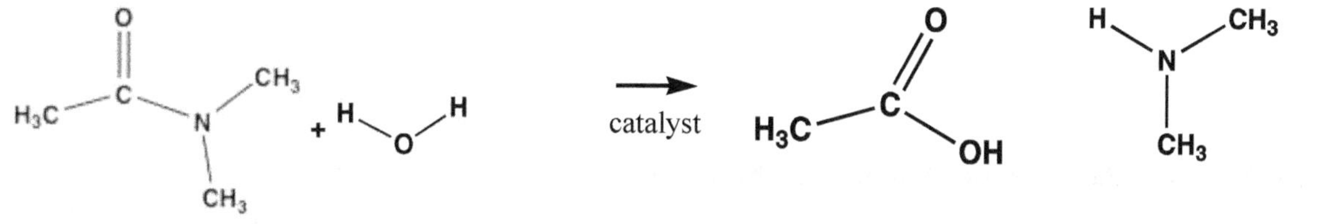

Critical Thinking Questions (cont'd.)

12. Draw an **X** through the bond in each reactant in Model 4 that is *broken*. *Circle* the H and OH groups from water that appear in the products. Which type of bond breaks in the organic reactants?

13. Does the carbonyl group itself react or change in these processes? (*Circle one.*) Yes/No

14. Describe how the hydrolysis of an *amide* fits the description of a "carbonyl group substitution reaction."

Summary (*Confer as a group.*)

Reactants needed to make an ester: ____________________ + ____________________

Reactants needed to make an amide: ____________________ + ____________________

Products of ester hydrolysis: ____________________ + ____________________

Products of amide hydrolysis: ____________________ + ____________________

Exercises (*Note:* For each reaction, assume appropriate catalysts are present.)

1. Write out the reaction to form an ester using propanoic acid and ethanol. Name the product.

2. Write out the reaction to form an amide using propanoic acid and ethylamine. Name the product.

3. Write out the hydrolysis reaction of ethyl butanoate. Name the products.

4. Write out the hydrolysis reaction of N-ethyl butanamide. Name the products.

5. Write out the reaction forming an ester from ethanedioic acid (oxalic acid) and 1, 2-ethanediol.

6. Write out the reaction forming an amide from propanedioic acid and 1, 2-diaminoethane.

7. What characteristic of the reactants in Exercises 5 and 6 could allow them to form polymers?

8. Be sure you *understand* the answers to the ***Critical Thinking Questions*** and ***Exercises*** in this activity. *Ask more questions* until you are confident in your answers.

9. Read the corresponding sections and work the suggested problems in the textbook.

8

REACTIONS OF AMINO ACIDS

How are organic reactions applied to amino acids?

Learning Objectives:

- Understand and apply amide formation and hydrolysis using amino acids.
- Relate reactions of carboxylic acids to those of amino acids.
- Describe protein primary structure characteristics.
- Determine names for simple peptides.

Prerequisite Concepts:

- Functional group recognition
- Substitution reactions
- Hydrolysis reactions
- Names of carboxylic acids, amines, amides
- Isomers
- Polymers

Model 1. Amide Formation

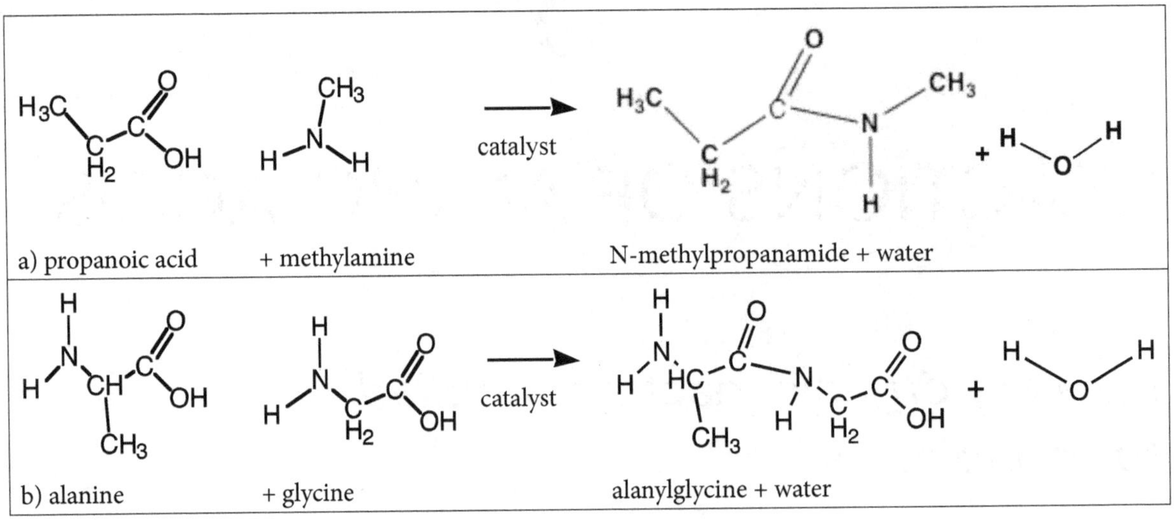

catalyst

a) propanoic acid + methylamine → N-methylpropanamide + water

catalyst

b) alanine + glycine → alanylglycine + water

CRITICAL THINKING QUESTIONS

1. Draw an **X** through the bonds that are *broken* in the reactants of Model 1. *Circle* the H and OH in the reactants that are the sources of H_2O in the product. Draw a *box* around the *amide* bond in each product.

2. Does the carbonyl group itself react or change in these processes? (*Circle one.*) Yes/No

3. *As a group,* determine how the formation of an amide fits the description of a "carbonyl group substitution reaction."

Information, Part I: Amides in Proteins

Heat or catalysts are required for the formation of amides whether in a lab or physiologically. Experimentally, e.g., in a procedure for preparation of nylon, specific catalysts or reactive acid derivatives such as acyl chlorides or acid anhydrides are used in place of carboxylic acids.

Alanine and glycine are two of the twenty naturally occurring difunctional molecules called α-amino acids used to make proteins. When the amide bond forms between two amino acids, it is termed a **peptide** bond since peptides are fragments of proteins. Proteins known as **enzymes** are used physiologically to catalyze amide bond formation and nearly all other biochemical processes.

Model 2. Hydrolysis of Amides

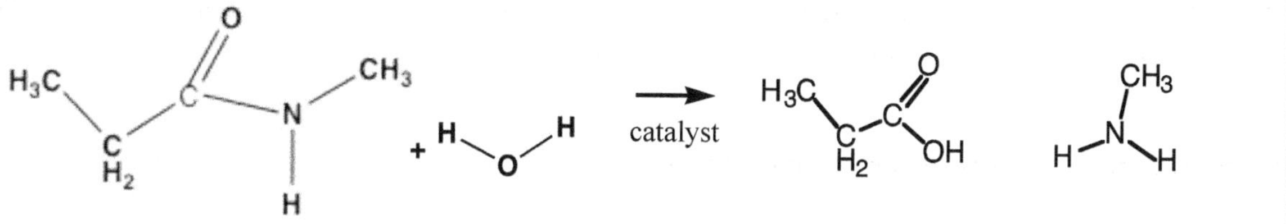

a) N-methylpropanamide + water	catalyst	propanoic acid	+ methylamine
b) alanylglycine + water	catalyst	alanine	+ glycine

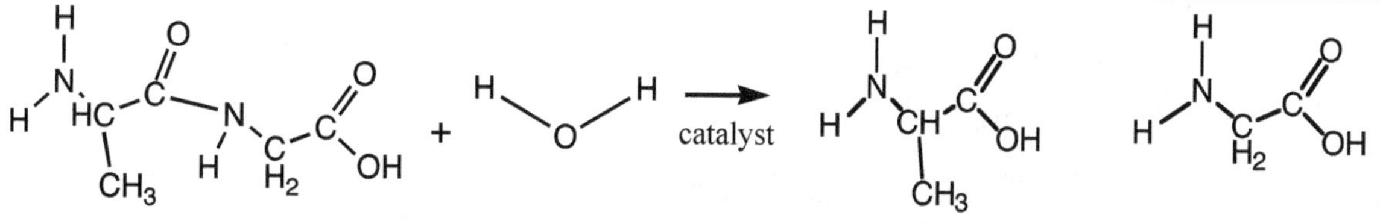

Critical Thinking Questions (cont'd.)

4. Draw an **X** through the bond in each reactant in Model 2 that is broken. *Circle* the H and OH groups from water that appear in the products. Which type of bond breaks in the organic reactants?

5. Does the carbonyl group itself react or change in these processes? (*Circle one.*) Yes/No

6. *As a group,* determine how the hydrolysis of an *amide* fits the description of a "carbonyl group substitution reaction."

Information, Part II: Hydrolysis in Proteins

Hydrolysis of amide bonds is required for the digestion of proteins. This occurs physiologically in the acidic media of the stomach using enzymes, the biological catalysts.

Model 3. Creating Primary Structure in Proteins

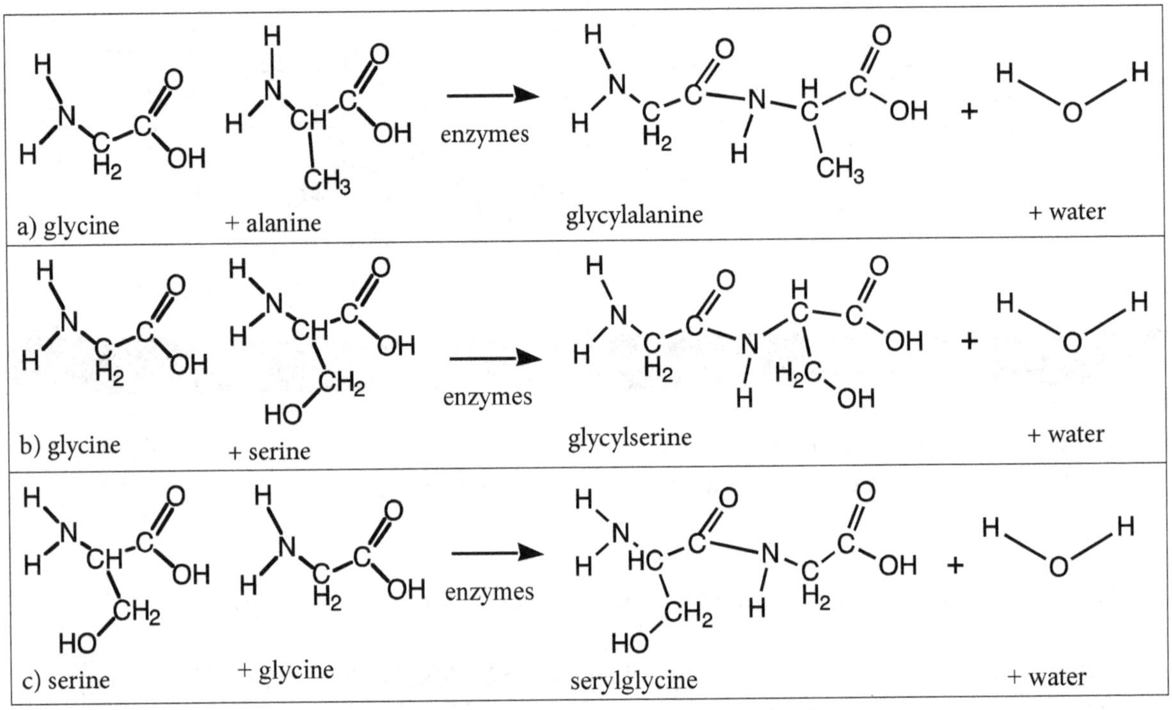

a) glycine	+ alanine	enzymes	glycylalanine	+ water
b) glycine	+ serine	enzymes	glycylserine	+ water
c) serine	+ glycine	enzymes	serylglycine	+ water

CRITICAL THINKING QUESTIONS (CONT'D.)

7. Make a *box* around the *amide* bond in each product of Model 3.

8. Compare the products shown in Model 1b to those in Model 3a. How are these products

   a. alike? b. different?

9. *As a group*, compare the products shown in Model 3b to those in 3c. How are these products

   a. alike? b. different?

10. What is the relationship between the products of Model 1b and 3a?

Same compound   Isomers   Unrelated

11. What is the relationship between the products of Model 3b and 3c?

Same compound   Isomers   Unrelated

12. *As a group,* note the names of the products in Model 3. These pairs of amino acids joined by an amide bond, known as dipeptides, are named by listing *first* the amino acid whose $-NH_2$ / $-COOH$ (*circle one*) group is unreacted and by changing the ending of its name to _____ . The *second* part of the name comes from the amino acid whose $-NH_2$ / $-COOH$ (*circle one*) group is unreacted and by *not* changing the ending.

13. For the dipeptides in Models 1 and 3, fill in the table below. For the shorter names, use the common three-letter abbreviations for the amino acids (*See* e.g., Table 9.1, p. 69).

Dipeptide	N-terminal amino acid	C-terminal amino acid	Shorter Name
alanylglycine	alanine	glycine	Ala-Gly
glycylalanine			
glycylserine			
serylglycine			

Information, Part III: Protein Structure

The order or sequence of amino acids in a protein describes the **primary structure,** which is held together by covalent peptide bonds. The amino acids are listed always starting with the N-terminal end. The **function** of a protein is very dependent on the overall **structure**. The three-dimensional shape of a protein is a combined result of the specific ordering of amino acids in the sequence, and the interactions between amino acids in different parts of the protein.

Exercises

1. For the dipeptide Phe-Glu, _____ is the N-terminal amino acid, and _____ is the C-terminal amino acid.

2. For the tetrapeptide Tyr-Thr-Cys-Asn, _____ is the N-terminal amino acid, and _____ is the C-terminal amino acid.

(*Note:* For the following reactions, assume appropriate catalysts are present.)

3. Write out the reaction to form an amide from glycine and the amino acid cysteine, H_2N-$CH(CH_2SH)$-COOH. What would it be named? (*See CTQ 12.*)

4. Write out the reaction to form a *different* amide formed from glycine and cysteine. What would it be named? ____________________

5. What characteristic of the reactants in Exercises 3 and 4 could allow them to form polymers?

6. Draw the structure of serylalanine. Write out the reaction to show the products of its hydrolysis.

7. What would be the relationship between serylalanine and alanylserine?

8. Draw the structure of alanylcysteine. Write out the reaction to show the products of its hydrolysis.

9. What would be the relationship between alanyserylcysteine and serylalanylcysteine?

10. Be sure you *understand* the answers to the ***Critical Thinking Questions*** and ***Exercises*** in this activity. *Ask more questions* until you are confident in your answers.

11. Read the corresponding sections and work the suggested problems in the textbook.

9

INTERACTIONS BETWEEN AMINO ACIDS IN PROTEINS

How do amino acid side chains influence protein 3° and 4° structure?

Learning Objectives:

- Describe the characteristics of amino acid side chains.
- Explain pH affects on side chains.
- Explain and predict interactions between amino acid side chains.
- Describe the similarities and differences between 3° and 4° protein structure influences.

Prerequisite Concepts:

- Functional group identification
- Protein primary structure
- Structures of alpha amino acids, side chains
- Peptides, peptide bonds
- Acid, base definitions
- Intermolecular forces

Critical Thinking Questions

(Circle those that apply in CTQs 1–5.)

1. Which kinds of functional groups are found in the side chains of the *nonpolar* amino acids?

   alkane alcohol aromatic amide amine carboxylic acid thiol

   alkene aldehyde ketone ether ester

2. Which kinds of functional groups are found in the side chains of the *polar* amino acids?

   alkane alcohol aromatic amide amine carboxylic acid thiol

   alkene aldehyde ketone ether ester

3. Which kinds of functional groups are found in the side chains of the *acidic* amino acids?

   alkane alcohol aromatic amide amine carboxylic acid thiol

   alkene aldehyde ketone ether ester

4. Which kinds of functional groups are found in the side chains of the *basic* amino acids?

   alkane alcohol aromatic amide amine carboxylic acid thiol

   alkene aldehyde ketone ether ester

5. *As a group*, determine which kinds of bonds are found in the side chain functional groups for these types of amino acids.

   polar amino acids C–C C–H C–O C–N O–H N–H

   nonpolar amino acids C–C C–H C–O C–N O–H N–H

6. Which kinds of intermolecular attractions are present for these side chain functional groups? Mark an **X** for attractions that apply for the group.

	alkane	alcohol	aromatic	amide	amine	carboxylic acid	thiol
London forces							
Dipole-dipole							
H-bonds							

7. Which amino acids have side chains that include these functional groups? List them below. (*Note:* Three-letter abbreviations for names are sufficient.)

alkane	amide
alcohol	carboxylic acid
aromatic	amine
thiol	thioether

Table 9.1. Amino Acids: Structures and Abbreviations
(*Note:* Structures are shown in fully ionized forms.)

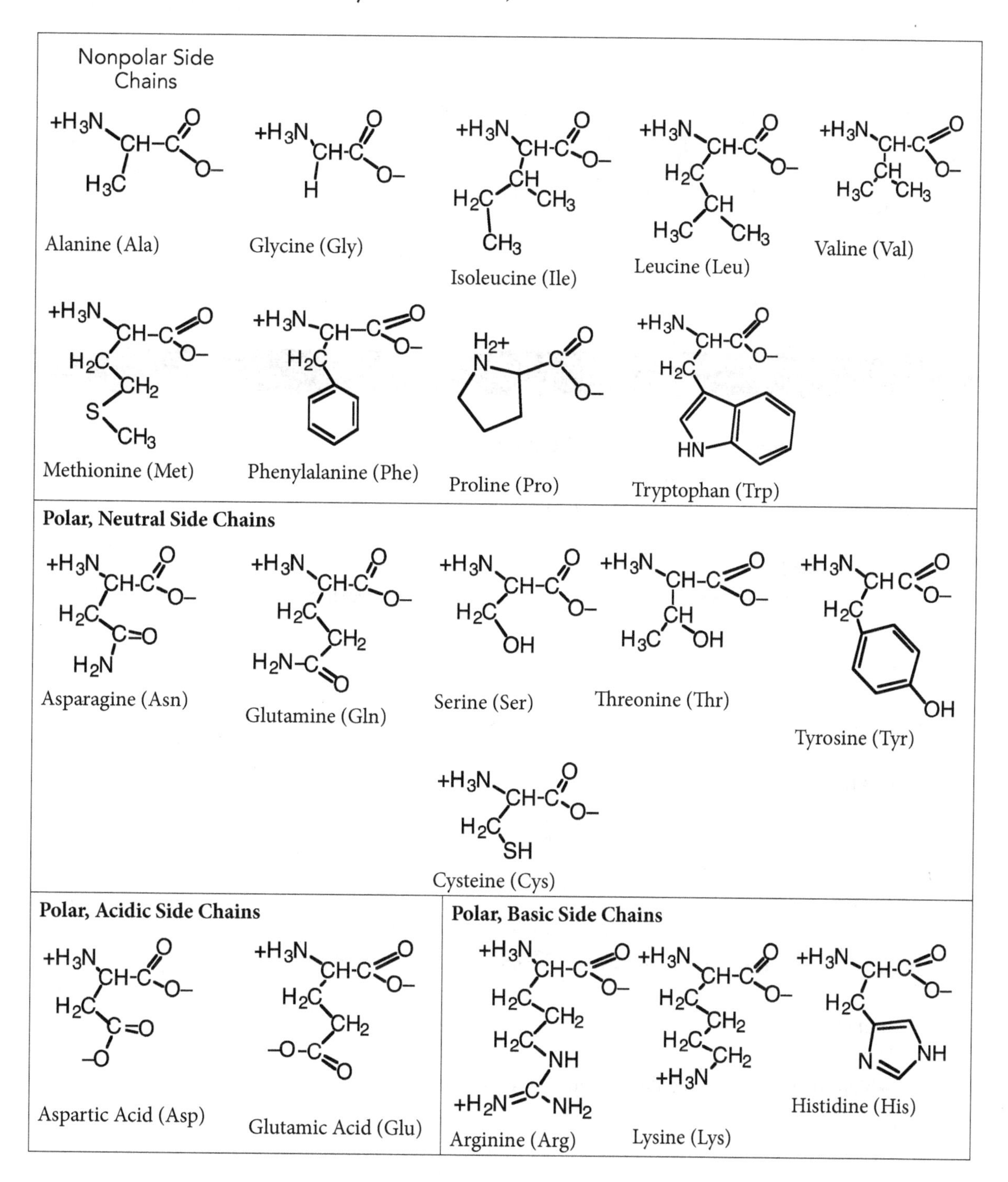

Model 1. Ionization of an Amino Acid

Lower pH: higher $[H_3O^+]$ more acidic			Higher pH: lower $[H_3O^+]$ more basic
(a)	(b)	(c)	(d)
Net charge =	Net charge =	Net charge =	Net charge =

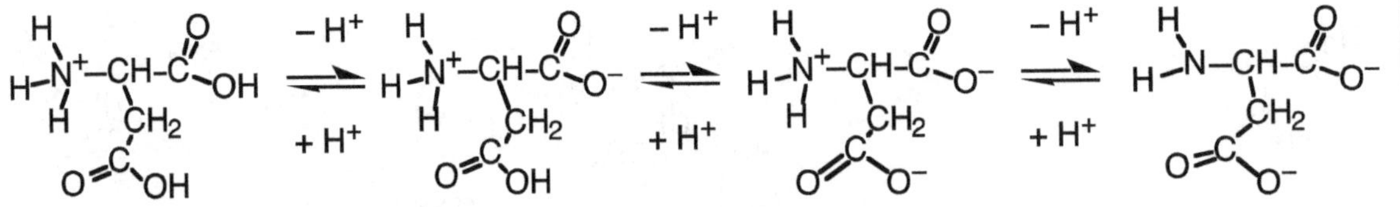

(a) ⇌ (b) ⇌ (c) ⇌ (d), each step $-H^+$ forward and $+H^+$ back.

Critical Thinking Questions (cont'd.)

8. Fill in the *net charge* for each form of the amino acid shown in Model 1. *Circle* the ionizable hydrogen atoms in each form shown, i.e., those that can be lost by ionization. *Compare your answers with those of your group or other groups nearby.*

(*Circle the word that makes each statement correct in CTQs 9–11*)

9. a. As the molecule *loses* H^+ ions, the net charge becomes increasingly

   positive negative

   b. As the molecule *gains* H^+ ions, the net charge becomes increasingly

   positive negative

10. a. At which pH does the molecule have the *most* hydrogen atoms?

    higher lower

    b. At which pH does the molecule have the *fewest* hydrogen atoms?

    higher lower

11. At which pH will the molecule in Model 1 be most *attracted* to

    a. *cationic* (positively charged) species? (e.g. Mg^{2+})

    lower pH higher pH

    b. *anionic* (negatively charged) species? (e.g. OH^-, NO_3^-, SO_4^{2-}, CO_3^{2-})

    lower pH higher pH

12. Identify the amino acid in Model 1.

Model 2. pH Effects on Amino Acids in Proteins

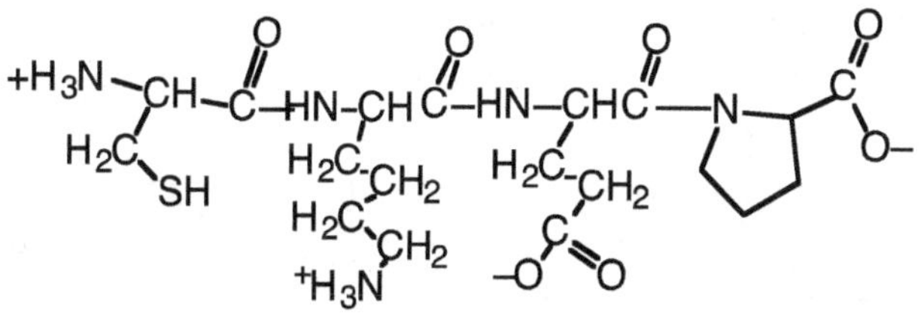

A tetrapeptide: **Cys-Lys-Glu-Pro**
(Shown with charges that would occur at physiological pH.)

Critical Thinking Questions (cont'd.)

13. Which form of the tetrapeptide from Model 2 shown below will predominate when the solution pH is lower (more acidic, high $[H_3O^+]$) or higher (more basic, low $[H_3O^+]$)? *(Circle one for each structure.)*

occurs at *lower pH / higher pH*	*lower pH / higher pH*
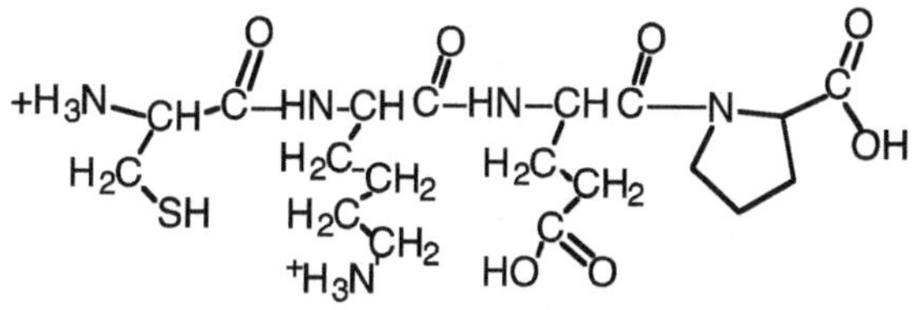	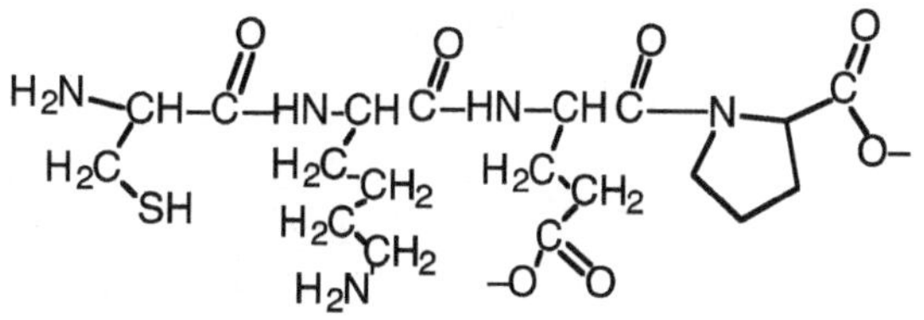

14. With which amino acid side chains from *Table 1* will the amino acids in the peptide of *Model 2* have *ionic* attractions? (*Note:* Some will have no ionic attractions. To clarify: Decide whether a Cys side chain would have ionic attractions to other amino acid side chains. If yes, which? Decide whether a Lys side chain would have ionic attractions to other amino acid side chains. If yes, which?)

Amino Acid	Cys	Lys	Glu	Pro
Attracted to these at physiological pH:				

15. *As a group*, determine which of these ionic attractions would be affected by a change in solution pH?

None *Most* *All*

Explain your answer.

Information, Part I: Side Chain Interactions

When the ionized side chains of amino acids in proteins attract one another, **salt bridges** can result. These attractions are dependent on the solution pH conditions. Only certain amino acids have ionizable side chains and can participate in salt bridging interactions.

Other interactions between amino acid side chains occur in proteins, including **hydrogen bonding** and hydrophobic interactions. When amino acid side chains cluster together to repel or exclude water, it is described as **hydrophobic interactions** (or repulsions or attractions).

Critical Thinking Questions (cont'd.)

16. Which amino acids have side chain functional groups that can participate in *hydrogen bonding*? (*Note:* Remember which groups are required for H-bonding: O–H or N–H. Look for 5–7 amino acids.)

17. Molecules that are attracted to water are described as:
    hydrophilic hydrophobic

    Molecules that repel water are described as:
    hydrophilic hydrophobic

18. *As a group*, determine and *circle* the amino acid side chain type(s) for each description.

Category	Side Chain Type			
Participate in salt bridges	Nonpolar	Polar, Neutral	Polar, Acidic	Polar, Basic
Participate in hydrogen bonding	Nonpolar	Polar, Neutral	Polar, Acidic	Polar, Basic
Hydrophobic	Nonpolar	Polar, Neutral	Polar, Acidic	Polar, Basic
Hydrophilic	Nonpolar	Polar, Neutral	Polar, Acidic	Polar, Basic

19. Only one amino acid has a thiol group that can oxidize to make a covalent *disulfide linkage* between two amino acid side chains. Which amino acid is this?

20. Fill in the chart using information gathered to this point.

Side Chain Interaction Summary

Interaction	Covalent or Noncovalent?	Side Chain *Type(s)* Involved
Salt Bridges		
Hydrogen Bonding		
Hydrophobic Interactions		
Disulfide linkages		

Information

The different levels of protein structure are dependent on covalent and noncovalent interactions. Protein **tertiary** (3°) structure is stabilized by the interactions between ***amino acid side chains***, both noncovalent and covalent. Protein **quaternary** (4°) structure is stabilized by the same noncovalent and covalent interactions between side chains, but the interactions occur between two or more polypeptide chains.

Exercises

1. Determine the net charge on each form of the amino acid shown. Indicate at which relative pH each form would be most prevalent. Which amino acid is shown? ____________

(a) net charge: ____	(b) net charge: ____	(c) net charge: ____
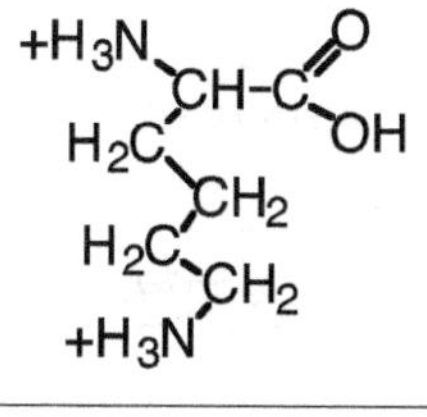 $+H_3N$, CH-C, O, OH H_2C CH_2 H_2C CH_2 $+H_3N$	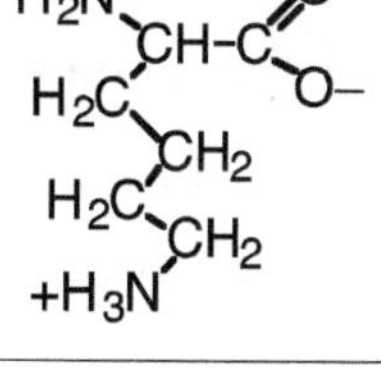 H_2N, CH-C, O, O– H_2C CH_2 H_2C CH_2 $+H_3N$	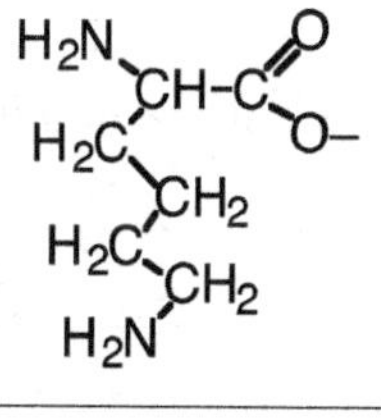 H_2N, CH-C, O, O– H_2C CH_2 H_2C CH_2 H_2N
Circle the pH where prevalent: lower / middle / higher	lower / middle / higher	lower / middle / higher

2. Referring to the molecule in Exercise 1, *circle* the correct term, and put the correct number in the blank.

   Structure (a) is converted to (b) by *gain / loss* of ___ H^+ ions.

   Structure (b) is converted to (c) by *gain / loss* of ___ H^+ ions.

   Structure (a) is converted to (c) by *gain / loss* of ___ H^+ ions.

   Structure (c) is converted to (a) by *gain / loss* of ___ H^+ ions.

3. For the tetrapeptide Arg-Asp-Phe-Gln shown below as it would occur at physiological pH, which amino acid side chains would have *ionic* attractions (salt bridges) with the side chains present?

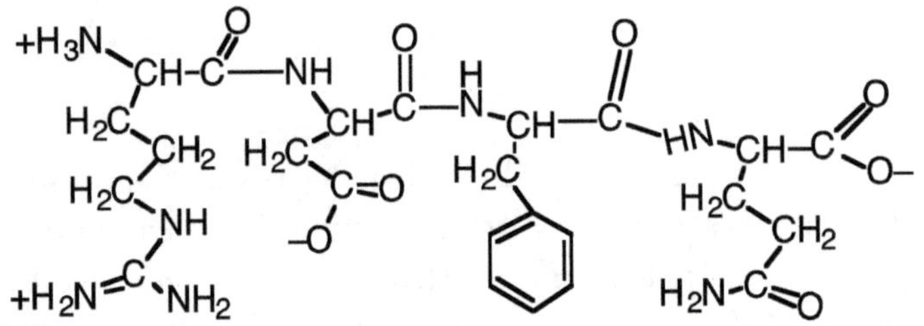

Amino Acid	Arg	Asp	Phe	Gln
Attracted to these at physiological pH:				

4. Draw the structures of serine and tyrosine, including any possible H-bonding interaction between their side chains.

5. Which side chain interactions are involved in these levels of protein structure?

3° Structure	4° Structure

6. What is the major difference between 3° and 4° structure?

7. Find one or more pairs of amino acids with *side chains* that would be expected to have each of these interactions:

   a Salt bridges

   b. H bonds

   c. Hydrophobic

   d. Disulfide

8. Be sure you *understand* the answers to the ***Critical Thinking Questions*** and ***Exercises*** in this activity. *Ask more questions* until you are confident in your answers.

9. Read the corresponding sections and work the suggested problems in the textbook.

10

STRUCTURES OF CARBOHYDRATES

What are important structural features for polyhydroxy-aldehydes and ketones?

Learning Objectives:

- Use numbering system in monosaccharides.
- Recognize chirality in carbohydrates.
- Explain intramolecular hemiacetal formation.
- Distinguish α and β forms of sugar anomers.

Prerequisite Concepts:

- Hemiacetal definition
- Bond polarity (**δ**+, **δ**–) in alcohols and carbonyls
- Chirality, stereoisomers, enantiomers, diastereomers
- Aldose, ketose, monosaccharide
- Identification of D and L sugars

Model 1. Intramolecular Attractions in a Monosaccharide

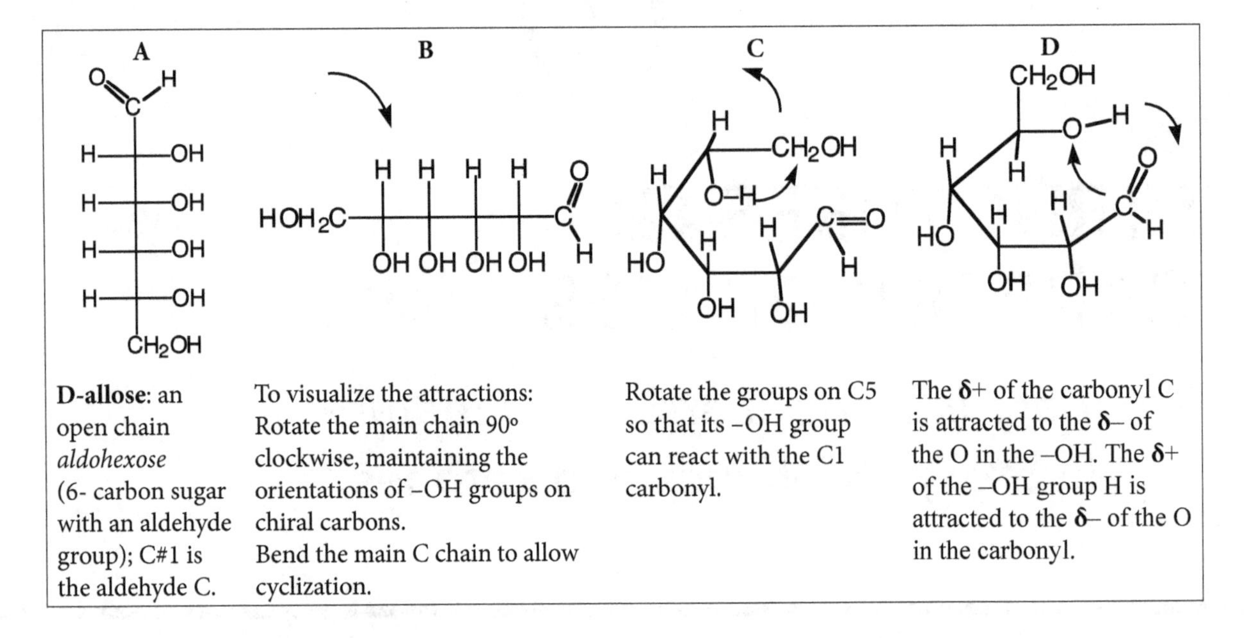

Critical Thinking Questions

1. Write in the numbers by the 6 carbons of D-allose in Model 1A just as you would for numbering an aldehyde. Number the carbons in Model 1B, C, and D in the same manner.

2. In Model 1D, mark the **δ**+ and **δ**– on the polar bonds of the (a) carbonyl and (b) C5 –OH groups.

3. Note the attractions between polar bonds in Model 1D.

   a. The (*choose one*) **δ**+ / **δ**– charge on the carbonyl C is attracted to the (*choose one*) **δ**+ / **δ**– charge of the O in the OH of carbon 5.

   b. The (*choose one*) **δ**+ / **δ**– of the H in the carbon 5 –OH group is attracted to the (*choose one*) **δ**+ / **δ**– of the O in the carbonyl.

4. Note the orientations of the OH groups on the structure in 1A.

   a. OH groups on the *right* in 1A are pointed (*choose one*) *down* / *up* in B, C, and D.

   b. If OH groups were found on the *left* in 1A, they would be pointed (*choose one*) *down* / *up* in B, C, and D.

Information, I: Representing Chirality in Carbohydrates

Carbohydrates are **polyhydroxy aldehydes** and **ketones**. The simplest ones, monosaccharides, have 3 to 6 carbons, one carbonyl group (the aldehyde or ketone group), and a hydroxyl (alcohol) group on all other carbons. The **Fischer projection** for monosaccharides, as shown in Model 1A, uses the *intersections of crossed lines to represent chiral carbons.*

CRITICAL THINKING QUESTIONS (CONT'D.)

5. How many chiral carbons are present in D-allose (Model 1A)?

   (*Circle one.*) 2 3 4 5 6

6. Which structure in Model 2 is the same as Model 1D? (*Circle one.*) 2B 2C

Model 2. Cyclization to Form an Intramolecular Hemiacetal in a Monosaccharide

A	B	C	D
β-D-allose	*rotating the C1-C2 bond*	*switches the carbonyl orientation*	**α-D-allose**

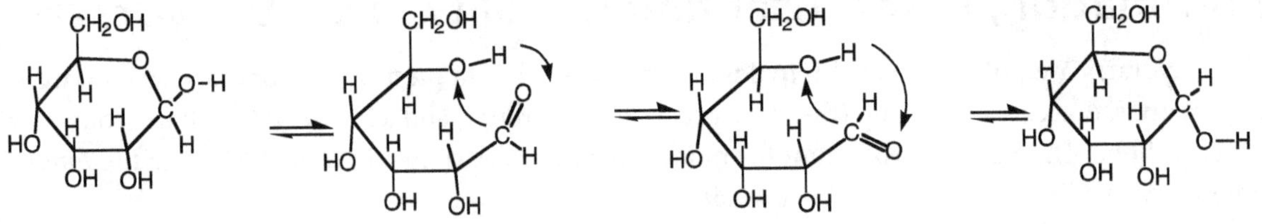

CRITICAL THINKING QUESTIONS (CONT'D.)

7. a. Number the carbons in Model 2A, 2B, and 2D using the same system as in Model 1.

   b. Put an *asterisk* (*) next to the carbon of the hemiacetals in Model 2A and 2D.

   c. Which number is this carbon? 2A: ________ 2D: ________

8. In Model 2A and 2D, *circle* the newly formed –OH group that is part of the hemiacetal.

9. *As a group*, determine the difference between the structures in Model 2A and 2D. Describe below the *relative* positions of the (i) Carbon #6 CH_2OH group and (ii) the newly formed -OH in the hemiacetal group for each structure.

   a. α–D-Allose:

   b. β-D-Allose:

10. *As a group*, determine how α–D-allose and β–D-allose are related. (*Circle those that apply.*)

    isomers   stereoisomers   enantiomers (mirror image pairs)

11. Which describes the formation of a hemiacetal? (*Circle one from each pair.*)

    reversible / forward only   covalent / non-covalent

Information, Part II: Hemiacetals and Anomers

Intramolecular hemiacetals form when an aldehyde or ketone reacts with an -OH group in the same molecule. Carbohydrates have the potential for intramolecular reactions whenever they have at least 5 carbons because the geometry allows for the requisite functional groups to approach close enough to react. Although the formation of a hemiacetal is reversible, those formed intramolecularly are very *stable* and are the predominant forms existing in solution.

Typical intramolecular hemiacetals formed by carbohydrates occur as 5- or 6-sided ***heterocycles*** with one oxygen atom, the oxygen of the hemiacetal's ether linkage.

The **anomeric** carbon is that which was originally the carbonyl of the aldehyde (or ketone) and becomes part of the hemiacetal (*marked with an asterisk in CTQ 7b*). The relative position of the –OH group of the hemiacetal determines whether the sugar is in the α or *β* form. Notice that the α and *β* forms are in ***equilibrium*** with the open chain forms (2B and 2C). Due to the stability of the hemiacetal, only a small fraction (<1%) of the sugar exists as the open chain form in solution.

The α and *β* forms of the sugar differ only in the orientation of groups on the anomeric carbon and are called α and *β* **anomers**.

Concept Check: *Find answers in the Information section above.*

12. How many carbons are needed for an intramolecular hemiacetal to form in carbohydrates?

13. Why are intramolecular hemiacetals the predominant forms in solution for carbohydrates?

14. Describe the heterocycles typical for carbohydrates in terms of numbers and types of atoms.

15. How can the anomeric C be identified?

16. Which two or three forms of a sugar are in equilibrium in solution?

17. How do the α and β forms of a sugar differ?

CRITICAL THINKING QUESTIONS (CONT'D.)

18. Which number is the anomeric carbon:

    In *α*–D-Allose? ____ In *β*–D-Allose? ____

19. Recall that to be chiral, a carbon must be *attached to four different groups*. Are the *anomeric* carbons in Model 2 chiral? (*Circle one in each pair.*)

    2A: Yes/No **2D**: Yes/No

Exercises

1. How many chiral carbons are present in each structure?

___ chiral C's	___ chiral C's	___ chiral C's	___ chiral C's
(a)	(b)	(c)	(d)

2. a) How are 1a and 1b related? (*Circle the terms that pertain.*)

   isomers stereoisomers enantiomers diastereomers

   b) How are 1b and 1c related? (*Circle the terms that pertain.*)

   isomers stereoisomers enantiomers diastereomers

3. a) Which sugars in Exercise 1 contain(s) a hemiacetal?

   (*Circle those that do.*) 1a 1b 1c 1d

   For those that do, put an asterisk (*) by the hemiacetal C.

   b) What is the number of the hemiacetal C in each case? ____________________

4. Which anomer type is shown in the sugars of Exercise 1?

   1a: **α** or **β** 1b: **α** or **β**

5. Which pairs of structures in Exercise 1 represent the open chain and cyclic hemiacetal forms of the same sugar? (*Hint:* Refer to Model 1 to recall the relationship between right–left vs. up–down orientations of OH groups.)

6. Which is the correct hemiacetal formed from the monosaccharides (i) and (ii) shown at the left? Circle the correct one. (*Hint:* Note the numbering of the carbons in the open chain form and the positions of each OH group.)

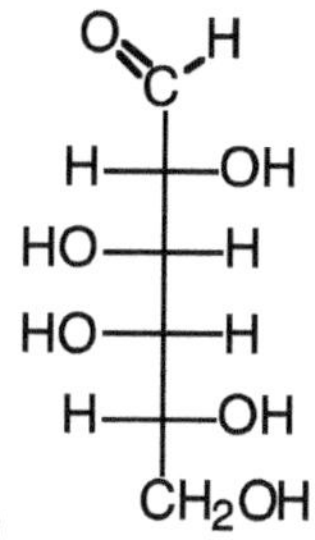 (i)	α/β 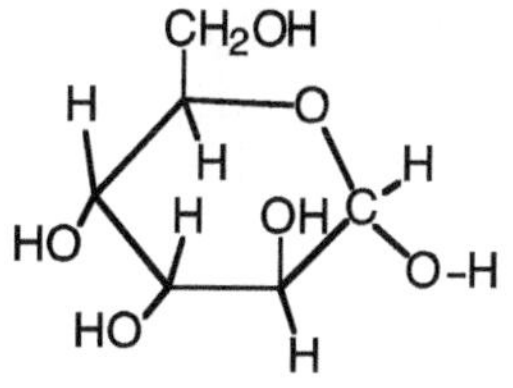	α/β 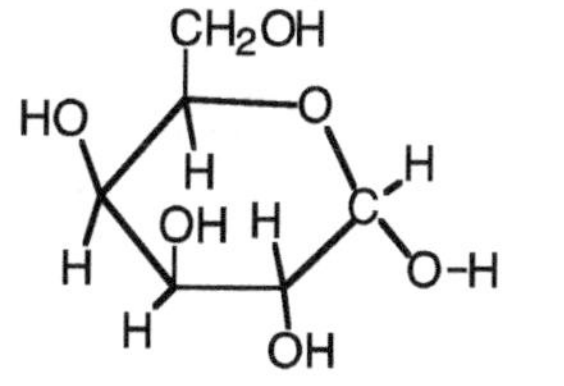	α/β 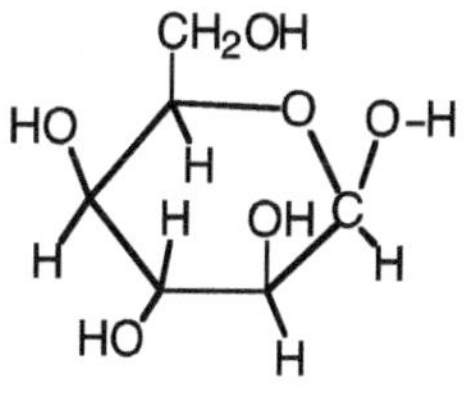
(ii) 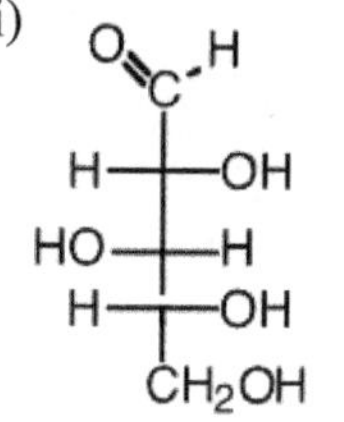	α/β 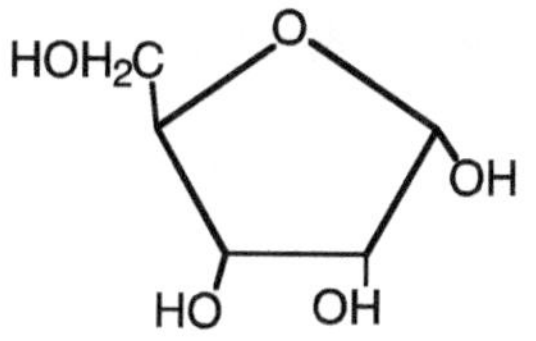	α/β 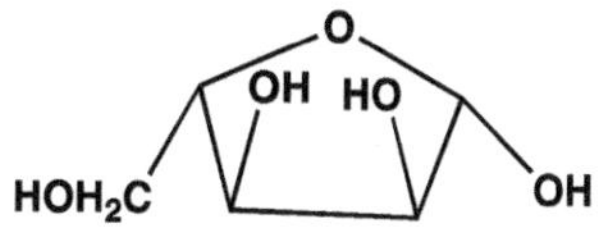	α/β 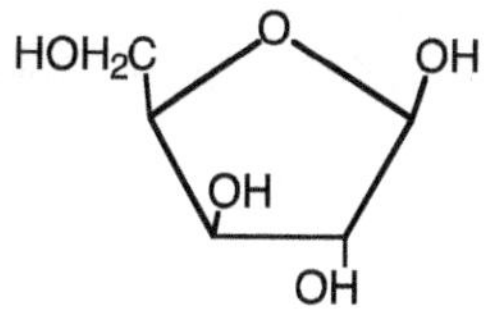

7. Label each sugar in Exercise 6 as an **α** or **β** anomer

8. Label each structure in Exercise 6 as D- or L-sugars.

9. Be sure you *understand* the answers to the ***Critical Thinking Questions*** and ***Exercises*** in this activity. *Ask more questions* until you are confident in your answers.

10. Read the corresponding sections and work the suggested problems in the textbook.

11

REACTIONS OF CARBOHYDRATES

Which reactions are important for polyhydroxy aldehydes and ketones?

Learning Objectives:

- Explain how oxidation of aldehydes applies to sugars.
- Describe how acetals are formed in carbohydrates.
- Identify acetals produced in reactions with carbohydrates.

Prerequisite Concepts:

- Oxidation, reduction
- Tollen's and Benedict's tests for sugars
- Hemiacetal, acetal definitions
- Anomers, α, β forms
- Numbering in carbohydrate structures

Model 1. Oxidation of a Monosaccharide

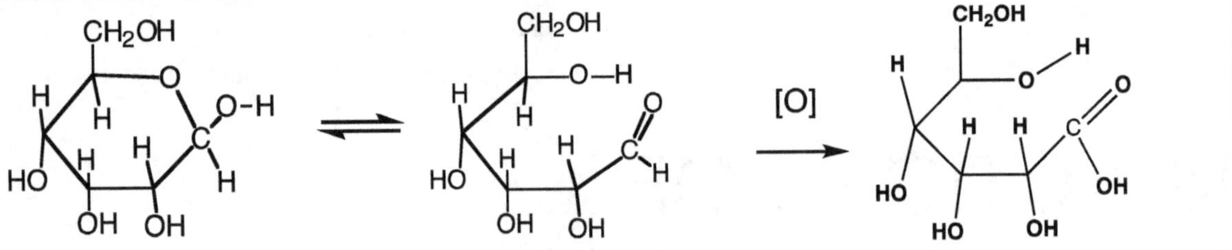

β-D-allose	(open chain aldehyde)	[O] →	oxidized product

CRITICAL THINKING QUESTIONS

1. Which form of allose undergoes oxidation? (*Circle one.*)
   hemiacetal   open chain

2. *As a group,* determine which group in allose becomes oxidized. (*Circle one.*)
   an –OH group   carbonyl of aldehyde

3. What new functional group results from this oxidation reaction?

4. Which describes the oxidation reaction? (*Circle one from each pair.*)
   reversible / forward only   covalent / non-covalent

Information, Part I: Reducing Sugars

Any monosaccharide in aqueous solution occurs as an open chain aldehyde *in equilibrium with* the cyclic α and β hemiacetal forms. When the aldehyde *oxidizes,* it allows another reagent to become *reduced.* In tests for aldehydes with Tollen's reagent, the Ag^+ becomes reduced to Ag metal, and in Benedict's reagent, Cu^{+2} ions become reduced to Cu^+ (in Cu_2O). In each of these tests, the aldehyde becomes *oxidized* to a carboxylic acid while the metal is *reduced.*

Similarly, in the presence of an oxidizer such as Fehling's reagent (also containing Cu^{+2} ions) used in qualitative tests for sugars, carbohydrates with an aldehyde group can be oxidized to carboxylic acid products while reducing the Cu^{+2} ions. Any sugar that can become oxidized while reducing another species is classified as a **reducing sugar**. Any sugar with a hemiacetal present qualifies as a reducing sugar because it is in equilibrium with an oxidizable open chain aldehyde form.

Concept Check: *Find answers in the Information section above to fill in the blanks for questions* 5–8.

5. In aqueous solutions, open chain forms of sugars are ______________________ with the cyclic α and β hemiacetal forms.

6. In the presence of Benedict's or Fehling's reagent, an aldehyde will be (*choose one*) oxidized / reduced to form a ________________ while the _____ (name the metal ion) is (*choose one*) oxidized / reduced to _____.

7. Any sugar that can become ____________ is a reducing sugar.

8. Any sugar that has a ______________ present is a reducing sugar.

CRITICAL THINKING QUESTIONS (CONT'D.)

9. a. *As a group*, determine whether β-D-allose is a reducing sugar. (*Circle one.*) Yes/No

   b. Would α-D-allose be a reducing sugar? (*Circle one.*) Yes/No

Model 2. Acetal Formation for a Monosaccharide

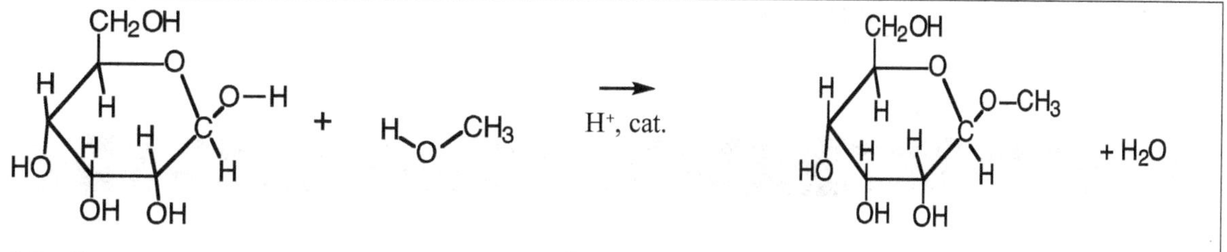

β-D-allose + methanol $\xrightarrow{H^+, \text{ cat.}}$ methyl β-glycoside of D-allose + H_2O

CRITICAL THINKING QUESTIONS (CONT'D.)

10. a. Which two functional entities are required to form an acetal?
    1. 2.

    b. What type of catalyst is shown for the acetal formation?

11. a. Mark the carbon in the center of the acetal group produced in the reaction of Model 2 with an asterisk (*).

    b. *Circle* the parts of the reactants that are removed to form water.

12. a. Number the carbons of the sugars. What number is the acetal carbon in Model 2?

    b. Is this the same as the anomeric carbon of the reactant? (*Circle one.*) Yes/No

13. *As a group*, decide whether or how the β designation for the acetal product of Model 2 is consistent with that used for hemiacetals. Explain your conclusion.

14. a. *As a group,* and based on the definition of a reducing sugar, decide whether the methyl β-glycoside of D-allose has the characteristics to be a reducing sugar. (*Circle one.*)

    Yes/No

    b. Why or why not?

Model 3. Acetal Formation Between Two Monosaccharides

α-D-allose	+	α-D-allose	H^+, cat. →	a disaccharide	+ H2O

Critical Thinking Questions (cont'd.)

15. Which *two* functional groups in the reactants are used to form the acetal?

16. *Circle* the groups in the reactants that are used to form water in the products.

17. a. Which number carbon of the left α-D-allose molecule is involved in the reaction?

    b. Which number carbon of the right α-D-allose molecule is involved in the reaction?

18. a) Which allose molecule contributes its anomeric carbon involved in the reaction? (*Choose one.*) left / right

    b) Which orientation is this anomeric carbon in the product? (*Choose one.*) α / β

19. *As a group,* using the numbers of the carbons reacted to make the acetal and the orientation of the anomeric carbon that reacted, complete the following description by using the answer from CTQ 18b in the first blank and the answers from CTQ 17 for the last two blanks.

    The product would be described as a(n) ___ - ___, ___ glycoside of allose.

Information, Part II: Acetals in Disaccharides

Acetals can form in carbohydrates by reaction of the hemiacetal with any alcohol-containing molecule, including other carbohydrates. When the acetal forms between two monosaccharides, a **disaccharide** results. Often, the resulting disaccharide is still a reducing sugar because the hemiacetal of only one sugar molecule reacts to form the acetal. Because acetals form in an acid catalyzed **condensation reaction** with water as the by-product, the resulting acetal is more stable than a hemiacetal. Acetals are only reconverted to aldehydes (or ketones) by acid catalyzed **hydrolysis,** the reverse of the reaction in Model 3. (Refer to your text for more details).

Concept Check: *Find answers in the Information section above.*

20. What groups react to form an acetal?

21. What linkage connects two sugars in a disaccharide?

22. When can a disaccharide be a reducing sugar?

23. How can an acetal in sugars be chemically broken?

Exercises

1.

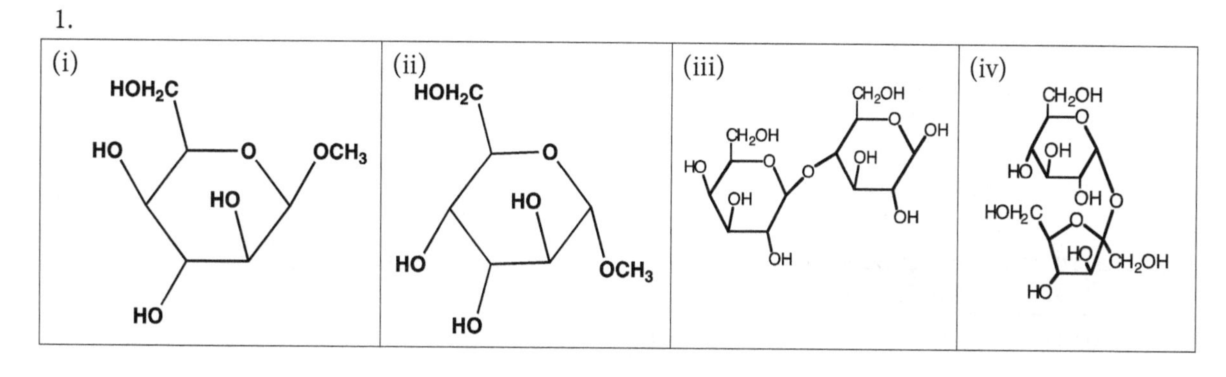

Which of the sugars in Exercise 1:

a. contain a hemiacetal? (*Circle those that do.*) i ii iii iv

b. are *reducing* sugars? (*Circle those that are.*) i ii iii iv

c. can be oxidized? (*Circle those that can be.*) i ii iii iv

2. Which anomer type is shown in the sugars of Exercise 1? (*Circle one from each section.*)

i: α or β iii: left sugar: α or β iv: top sugar: α or β

ii: α or β iii: right sugar: α or β iv: bottom sugar: α or β

3. Which sugars in Exercise 1 contain an acetal linkage? Draw a *box* around any acetal carbons in these structures. (*Circle those that have acetals.*) i ii iii iv

4. What are the numbers of the carbons of the sugars that are connected in Exercise 1iii and 1iv? (*Hint:* Note that the *lower* sugar in iv has two CH_2OH groups; the *right* CH_2OH is Carbon 1)

   sugar iii: _____ and _____ sugar iv: _____ and _____

5. Draw the product of oxidation of galactose. (*Hint*: Refer to Model 1 for an example.)

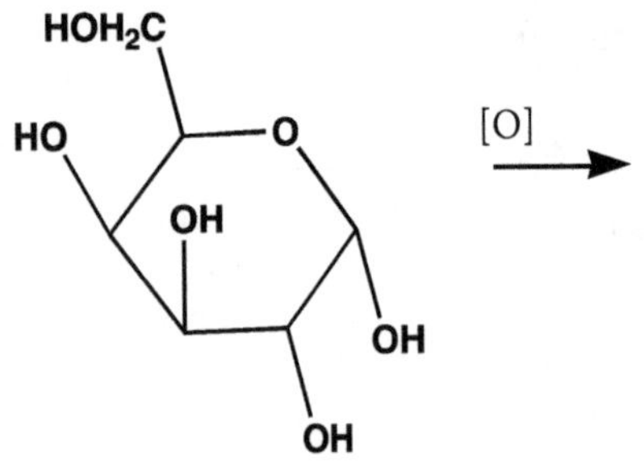

HOH₂C HO O OH OH OH [O] →

D-galactose

6. Draw the acetal resulting from reaction of galactose with ethanol.

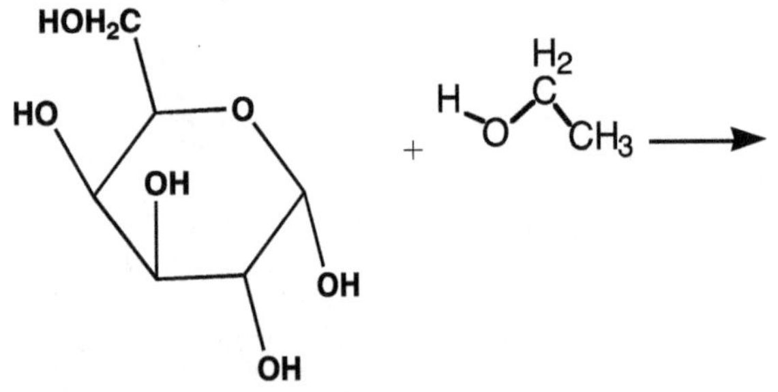

HOH₂C HO O OH OH OH + H–O–CH_2–CH_3 →

D-galactose

7. Describe the acetal product in Exercise 6. (*Hint:* See Model 2 and CTQ 19.)

   _______ - glycoside of _________________
   (α *or* β) (*alkyl group name*)

8. a. Can the product of Exercise 6 be oxidized? (*Circle one.*) Yes/No
   b. Is the product a reducing sugar? (*Circle one.*) Yes/No
   c. Why or why not?

9. Be sure you *understand* the answers to the ***Critical Thinking Questions*** and ***Exercises*** in this activity. *Ask more questions* until you are confident in your answers.

10. Read the corresponding sections and work the suggested problems in the textbook.

12

REACTIONS OF ESTERS IN LIPIDS

How are organic reactions applied to triglycerides?

Learning Objectives:

- Explain and apply esterification of carboxylic acids to triglycerides.
- Relate acid–base properties of carboxylic acids to their reactions.
- Explain the hydrolysis of esters in lipids.
- Identify the similarities between carboxylate and phosphate esters.

Prerequisite Concepts:

- Functional group recognition (e.g., esters, acids)
- Carbonyl group substitution reactions
- Acid–base properties of carboxylic acids
- Acid–base conjugate pairs
- Hydrolysis reactions
- Catalysts
- Names of carboxylic acids, esters

Model 1. Ester Formation, a Review and an Application

a)

propanoic acid + methanol $\xrightarrow{[H^+]}$ methylpropanoate + water

b)

$3\ CH_3(CH_2)_{10}COOH$ + glycerol $\xrightarrow{[H^+]}$ a triglyceride + 3 water

CRITICAL THINKING QUESTIONS

1. In Model 1, draw an X through the bonds that are broken in the reactants. *Circle* the H and OH in the reactants that are the sources of H_2O in the product. Draw a *box* around the *ester* bond(s) in each product.

2. Does the carbonyl group itself react or change in these processes? (Circle one.) Yes/No

3. Recall the description of a *carbonyl group substitution reaction* (*Note:* Refer to Activity 7, *Reactions of Carboxylic Acids*), and explain how the formation of these esters fits that description.

4. *As a group,* determine why the "triglyceride" product of Model 1b is described as a *triester.*

Information, Part I

One common type of lipids is triglycerides (triacylglycerols), the triesters of glycerol (1, 2, 3-propanetriol) with long chain carboxylic acids known as fatty acids. These lipids vary in the types of fatty acids included, depending on the sources from which they are isolated. They are formed in reactions controlled by enzymes, the proteins that serve as biological catalysts,

Model 2. Review of Acid–Base Reactions of Carboxylic Acids

a) CH_3CH_2COOH propanoic acid	+ NaOH	⟶ $CH_3CH_2COO^-$ + Na^+ + H_2O
b) CH_3CH_2COOH propanoic acid	+ HCl	⟶ *no reaction*
c) CH_3CH_2COOH propanoic acid	+ H_2O	⇌ $CH_3CH_2COO^-$ + H_3O^+
d) $CH_3CH_2COO^-$ propanoate	+ HCl	⟶ CH_3CH_2COOH + Cl^-
e) $CH_3CH_2COO^-$ propanoate	+ NaOH	⟶ *no reaction*

Critical Thinking Questions (cont'd.)

5. Draw a *circle* around each species in Model 2 that is an *acid.* Draw a *box* around each species that is a *base.* Draw *arrows* to show which pairs of compounds are acid–base *conjugate pairs.* The first one is done for you.

6. *As a group*, based on reactions shown in Model 2, complete the generalizations about the forms present for a carboxylic acid in different solution conditions. (*Circle the correct choice for each.*)

   a. In the presence of an acid (i.e., at lower pH), a carboxylic acid ________________.

   ionizes to form a carboxylate anion — remains protonated

   b. In the presence of a base (i.e., at higher pH), a carboxylic acid ________________.

   ionizes to form a carboxylate anion — remains protonated

   c. In water, a carboxylic acid ______________________________ its carboxylate ion.

   converts to — is in equilibrium with

   d. Carboxylate ions are re-converted to carboxylic acids by reaction with ________.

   acids — bases — water

Critical Thinking Questions (cont'd.)

7. Draw an **X** through the bond(s) in each reactant in Model 3 that is broken. *Circle* the –H and –OH groups from water that appear in the products for reactions (a) and (c).

8. Does the carbonyl group itself react or change in these processes? (*Circle one.*) Yes/No

9. Which of the products are the same in Model 3a and 3b?

10. Which of the products are the same in Model 3c and 3d?

Model 3 Hydrolysis of Esters

Reactants	Catalyst	Products
$H_3C-CH_2-C(=O)-O-CH_3$ + H–O–H methyl propanoate + water	(a) $[H^+]$ → *{in acid}*	$H_3C-CH_2-C(=O)-OH$ + $H-O-CH_3$ propanoic acid + methanol
	(b) NaOH → *{in base}*	$H_3C-CH_2-C(=O)-O^-$ Na^+ + $H-O-CH_3$ + H–O–H sodium propanoate + methanol + H–OH
$H_2C-O-C(=O)-R$ $CH-O-C(=O)-R$ $H_2C-O-C(=O)-R$ triglyceride + 3 water (*triacylglycerol*)	(c) $[H^+]$ → *{in acid}*	H_2C-OH, $CH-OH$, H_2C-OH + 3 $H-O-C(=O)-R$ glycerol + 3 carboxylic acids
	(d) NaOH → *{in base}*	H_2C-OH, $CH-OH$, H_2C-OH + 3 Na+ $^-O-C(=O)-R$ + H–O–H H–O–H H–O–H glycerol + 3 carboxylate salts + 3 water

11. What do the products of Model 3a and 3c have in common?

12. What do the products of Model 3b and 3d have in common?

13. *As a group,* determine why the products from Model 3a and 3b are different.

14. *As a group,* determine why the products from Model 3c and 3d are different.

Information, Part II: Lipid Hydrolysis

Hydrolysis reactions can occur at any ester bond in a molecule. Experimentally, all ester bonds would hydrolyze given sufficient time and a suitable catalyst. When a triglyceride of fatty acids is hydrolyzed using a base, the process is termed **saponification** because the resulting carboxylate anions, combined with metal cations, are salts known as **soaps.** The carboxylate salts of long chain carboxylic (fatty) acids act as soaps due to the presence of both *polar* (the ionic part) and *nonpolar* (the hydrocarbon chain) ends.

Physiologically, ester hydrolysis is controlled by enzymes, the biological catalysts that control the specific bonds that are broken. Using enzymes, reactions proceed under normal physiological conditions of temperature and pH.

Concept Check: *Find answers in the Information section above to fill in the blanks for CTQ 15–18.*

15. What is the process of hydrolysis of a triglyceride in the presence of a base called?

16. What product results from this reaction?

17. What characteristics do these products have that allow them to act as soaps?

18. How does hydrolysis take place physiologically?

Model 4. Phosphate Ester Formation

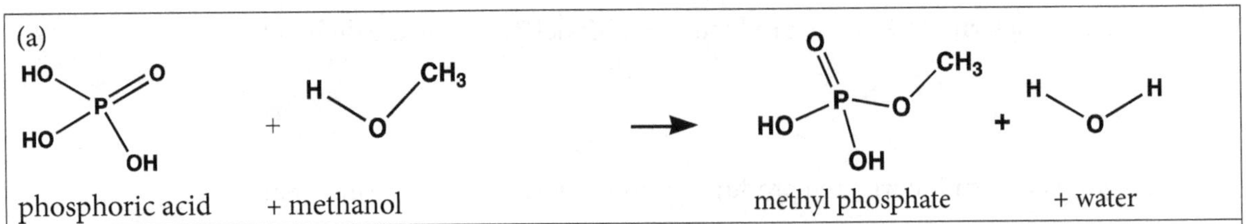

(a)

phosphoric acid + methanol → methyl phosphate + water

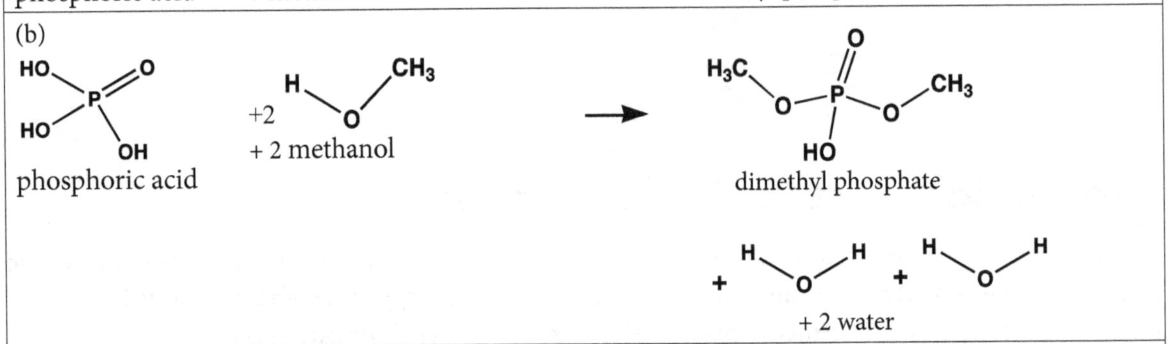

(b)

phosphoric acid + 2 methanol → dimethyl phosphate + 2 water

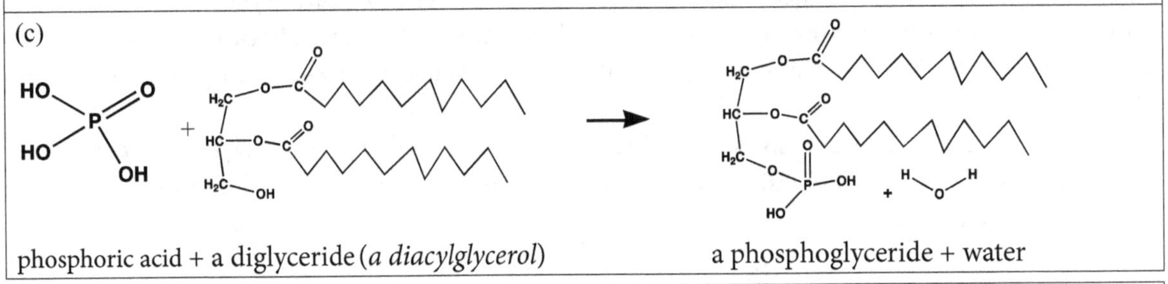

(c)

phosphoric acid + a diglyceride (*a diacylglycerol*) → a phosphoglyceride + water

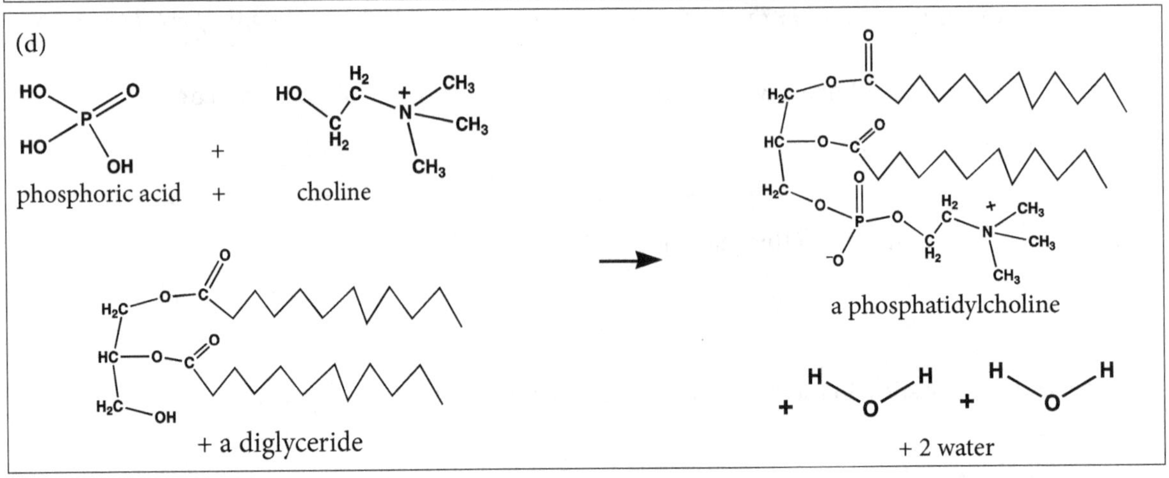

(d)

phosphoric acid + choline + a diglyceride → a phosphatidylcholine + 2 water

Critical Thinking Questions (cont'd.)

19. Draw an **X** through the bonds that are broken in the reactants of Model 4. *Circle* the H and OH in the reactants that are the sources of H_2O in the product. Draw a *box* around any *phosphate ester* bond(s) in each product in Model 4.

20. Which product(s) in Model 4 might be described as containing "*phosphodiester*" linkages?

21. *As a group*, describe how the formation of a phosphate ester is similar to a "*carbonyl group substitution reaction.*"

22. The product in the last reaction of Model 4 is shown with a negative charge on the phosphate group and has a positive charge on the choline fragment. How is this product similar to a soap? Compare answers among your group members or with a neighboring group.

Exercises

(*Note:* For each reaction, assume appropriate catalysts are present.)

1. Write the esterification reaction using butanoic acid and ethanol. Name the product.

2. Write the reaction forming a diester from 1, 2-ethanediol with 2 molecules of butanoic acid.

3. Write the reaction forming a triester from glycerol and 3 molecules of stearic acid ($CH_3(CH_2)_{16}COOH$).

4. Write the reactions forming:
   a. a phosphate ester from phosphoric acid and ethanol.

   b. a phosphodiester from phosphoric acid, ethanol, and the C5 –OH of ribose.

5. Write the hydrolysis products expected from propyl ethanoate ($CH_3CH_2CH_2$-OOC-CH_3) when reacted with:

   a. an acid catalyst (e.g. H_2SO_4): (*Hint*: Like Model 3.)

   b. a base catalyst (e.g., NaOH):

6. Draw the products expected from complete hydrolysis of phosphatidyl choline when reacted with:

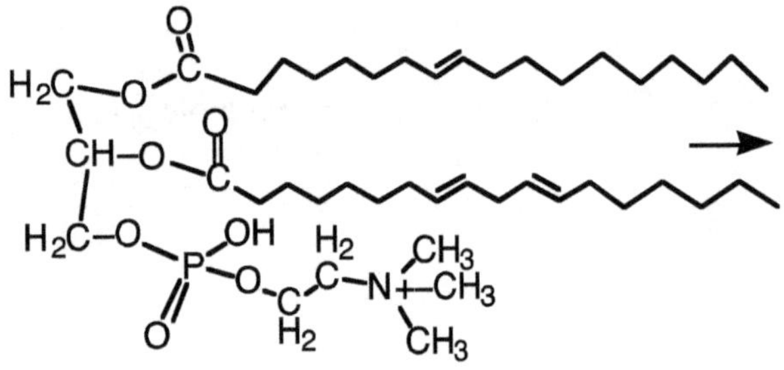

   a. an acid catalyst (e.g., HCl):

   b. a base catalyst (e.g., NaOH): →

7. Label any products from Exercise 6 that would be called soaps.

8. Be sure you *understand* the answers to the ***Critical Thinking Questions*** and ***Exercises*** in this activity. *Ask more questions* until you are confident in your answers.

9. Read the corresponding sections and work the suggested problems in the textbook.

13

ENZYMES

What are important features of biochemical catalysts?

Learning Objectives

- Recognize specificity types in enzyme-catalyzed processes.
- Understand enzyme inactivation and inhibition.
- Identify classes of enzymatic reactions.

Prerequisite Concepts

- Protein 1°, 2°, 3°, 4° structure
- Protein denaturation, hydrolysis
- Hydrolysis reactions of acetals, amides
- Organic reaction types
- Glucose structure
- Acetal linkages
- Di- and polysaccharides, compositions
- Oxidation–reduction reactions

Model 1. Carbohydrate Digestibility

Sugar	Type	Linkage Type(s)	Digestible By
Maltose	disaccharide	α-1, 4 glucoside	Humans
Amylose	polysaccharide	α-1, 4 glucosides	Humans
Amylopectin	polysaccharide	α-1, 4 and α-1,6-glucosides	Humans
Lactose	disaccharide	β-1, 4 galactoside with glucose	Many humans
Cellulose	polysaccharide	β-1, 4 glucosides	Microorganisms in digestive tracts of cows & termites

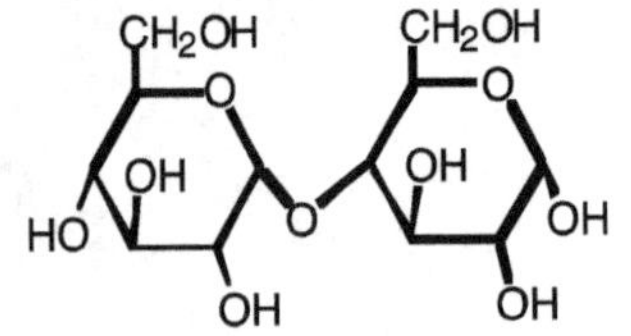

Maltose

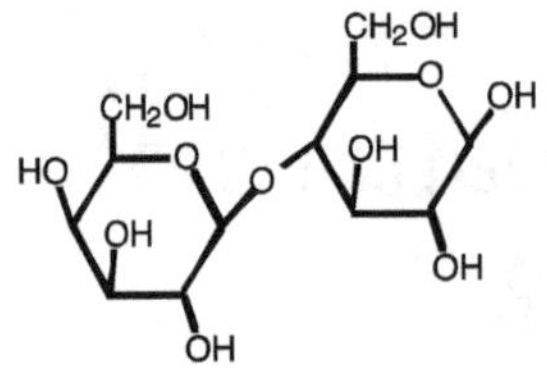

Lactose

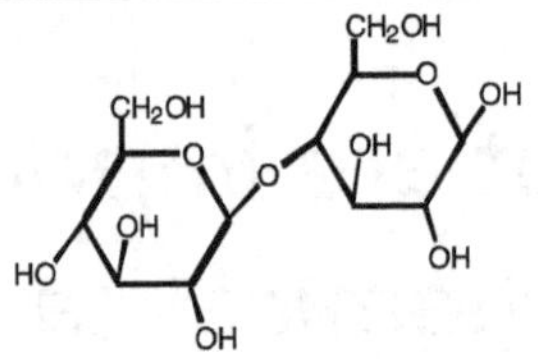

Cellobiose (disaccharide from cellulose)

CRITICAL THINKING QUESTIONS

1. What linkage is present that must be digested in the sugars of Model 1? (*Circle your choice.*)
   acetal amide ester hemiacetal

2. *As a group,* decide what seems to be a significant characteristic in determining which di- or polysaccharides are digestible by humans, i.e., what they have in common.

Information, Part I: Carbohydrate Digestion

Carbohydrate digestion requires hydrolysis of the linkages between sugars. In the laboratory, acid catalysts are required. Physiologically, carbohydrates are digested by acid in the stomach, and by enzymes, such as **α-amylase** in the saliva, and others in the intestines. Enzymes often have specificity for certain types of linkages or orientations in biomolecules. **α**-amylase hydrolyzes only **α**-glucosidic (acetal) linkages. **β**-glucosidic (acetal) linkages are not hydrolyzed by **α**-amylase, and are therefore less easily digested.

Critical Thinking Questions (cont'd.)

3. List the sugars from Model 1 that would be hydrolyzed during digestion by α-amylase.

4. Yogurt, prepared from milk curdled by bacteria, can be eaten by lactose-intolerant individuals without difficulty. *As a group,* explain why such individuals can tolerate this milk product.

Information, Part II: Activity and Inactivation of Enzymes

Enzymes catalyze reactions by binding to a reactant, called a **substrate,** for enzymatic processes; the binding occurs at a part of the enzyme called the **active site** where the catalysis will occur. A substrate is held in the active site by the same non-covalent interactions that stabilize protein structures. (*Note*: Refer to Activity 9, *Interactions Between Amino Acids in Proteins.*) A substrate must have the proper three-dimensional fit with the active site to bind to and be catalyzed by an enzyme. Changes in the structure of the active site or of the substrate result in changes in the catalytic activity of the enzyme.

Enzymes are proteins. Anything that affects protein *structure* affects enzyme *function* and, therefore, activity. **Denaturation** and **hydrolysis** disrupt or destroy the normal *structures* of proteins and, therefore, their *functions.*

Concept Check: *Find answers in the Information section above.*

5. What is the reactant in an enzymatic reaction called?

6. Where on the enzyme does catalysis occur?

7. Name three interactions that hold a substrate in the active site.

    1.

    2.

    3.

8. Anything that affects an enzyme's __________ also affects its ____________.

9. What types of processes disrupt enzyme structure?

Critical Thinking Questions (cont'd.)

10. List 6 (or more) agents that cause protein denaturation. (*Refer to your text for hints.*)

    1. 2. 3.

    4. 5. 6.

11. What is required for hydrolysis of proteins? (*Circle your choice[s].*)
    acid catalyst + heat enzymes base catalyst + heat any/all of these

12. Which levels of protein structure are affected by:

    a) Denaturation? 1° 2° 3° 4°

    b) Hydrolysis? 1° 2° 3° 4°

13. Preserving fresh fruits and vegetables involves processes that *inactivate* enzymes, which can cause deterioration of the food. For freezing, foods are first briefly subjected to boiling water; this process is known as "blanching." How does this aid in the preservation? (*Hint:* Think in terms of hydrolysis or denaturation.) *Discuss the answer among your group members.*

14. Some foods are preserved by pickling in vinegar solutions. How does this benefit the preservation?

Model 2. One Example of Enzyme Inhibition

acetylsalicylic acid + active cyclooxygenase → salicylic acid + inactive cyclooxygenase

The antipyretic (fever-reducing) effect of aspirin is due to the inhibition of cyclooxygenase, an enzyme involved in prostaglandin synthesis. The inactive enzyme cannot convert arachidonic acid to a prostaglandin.

Critical Thinking Question (cont'd.)

15. In becoming inactive, how is the enzyme cyclooxygenase likely altered by aspirin? (*Circle any that apply.*)

    covalently noncovalently change in structure change in protein shape

Information, Part III: More Types of Enzyme Inhibition

A change in structure can cause a change that **activates** or inactivates an enzyme's action. An inactivation or **inhibition** of an enzyme can occur when the active site is blocked by the reversible noncovalent binding of a molecule that is structurally similar to the normal substrate. This is called **competitive inhibition** because the inhibitor competes with the substrate for binding to the enzyme's active site. If an inhibition occurs by the binding of a molecule somewhere other than the active site, it is described as **noncompetitive inhibition.**

Concept Check: *Find answers in the Information section above.*

16. Which type of inhibition results when the active site is reversibly blocked by the binding of a molecule similar to the substrate?

17. Which type of inhibition results from the binding of a molecule somewhere other than the active site?

Critical Thinking Questions (cont'd.)

18. *As a group,* determine which inhibitor type of cyclooxygenase fits for acetyl salicylic acid: (*Circle one.*)

    reversible, competitive reversible, noncompetitive irreversible

19. *As a group,* determine which molecule is most likely to act as a competitive inhibitor for α-amylase? (*Circle one.*) (*Hint:* Refer to Information, Part I.)

    β-L-glucose α-D-glucose β-D-galactose

Model 3. Enzyme Classes

Oxidoreductases	**A**(*reduced*) + **B**(*oxidized*)	$\longrightarrow$	**A**(*oxidized*) + **B**(*reduced*)
Transferases	**A + B–C**	$\rightleftharpoons$	**A–B + C**
Hydrolases	**A–B + H_2O**	$\longrightarrow$	**A–OH + B–H**
Isomerases	**A**	$\rightleftharpoons$	**B**
Lyases	A(H)C=C(B)H + H_2O	$\rightleftharpoons$	A–C(HO)(H)–C(H)(H)–B
Ligases	**A + B + ATP**	$\longrightarrow$	**A–B + ADP + HPO_4^{-2} + H^+**

Critical Thinking Questions (cont'd.)

20. *As a group*, determine to which of the six enzyme classes these enzymes would belong.

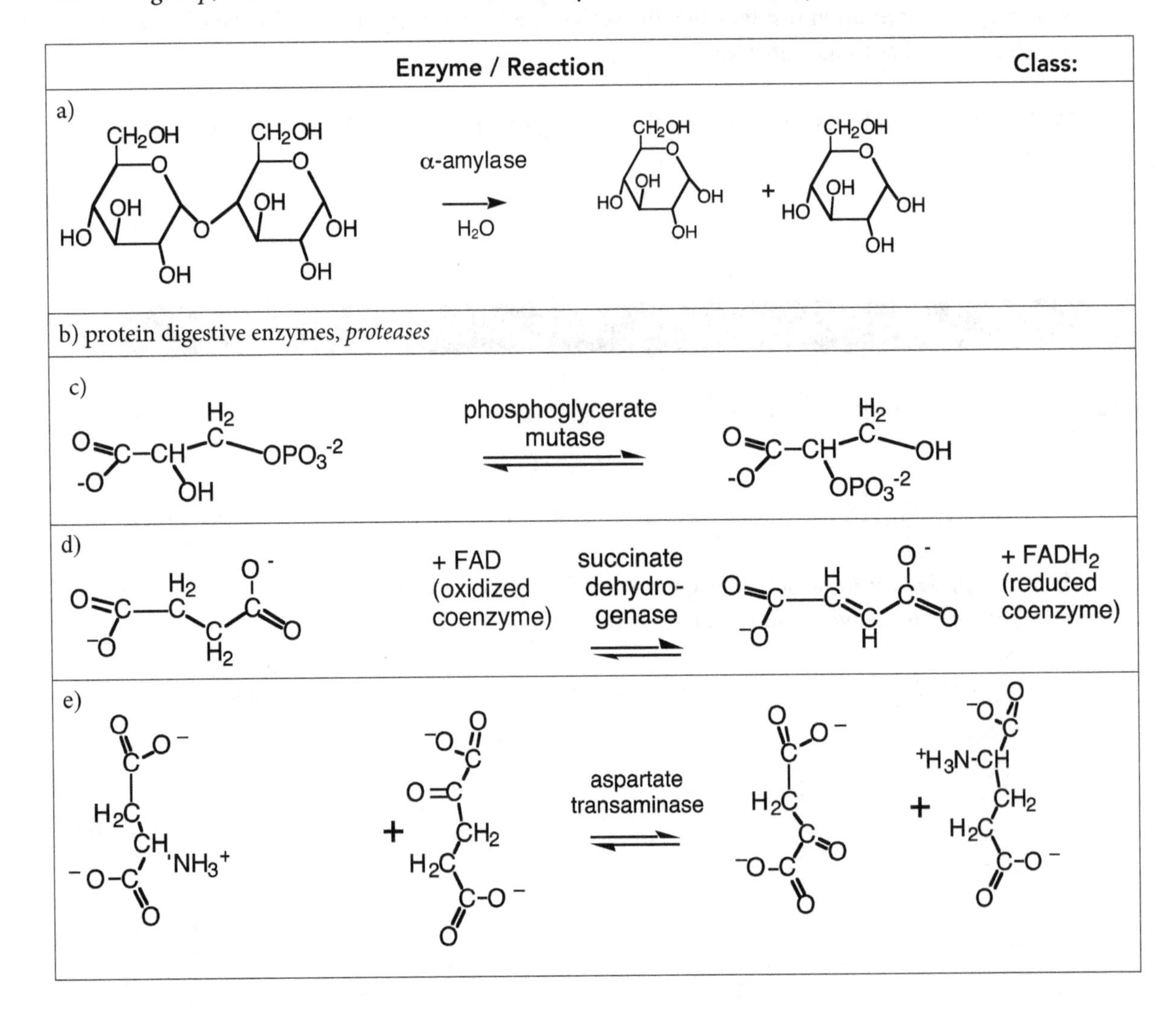

Exercises

1. **Beano** is a food enzyme dietary supplement that contains *alpha*-galactosidase derived from the mold *Aspergillis niger*. The enzyme breaks down the complex sugars from beans, peas, and other legumes that are high in small polysaccharides containing galactose bonded to glucose. These polysaccharides are digested by bacteria in the gut producing lactate, short chain fatty acids, and gaseous by-products including H_2, CO_2, and CH_4.

   a. To which class of enzymes would *alpha*-galactosidase belong?

   b. Which unusual type of linkage would likely be present in polysaccharides found in legumes? (*Circle one.*)

   α-glucosides β-glucosides α-galactosides β-galactosides

   c. What molecule might be a competitive inhibitor for *alpha*-galactosidase?

   d. Package information for the product suggests not to cook with Beano®. Why would that be important for its effectiveness?

2. Glucose is converted to glucose-1-phosphate in glycolysis.

   a. To which class of enzymes does hexokinase belong?

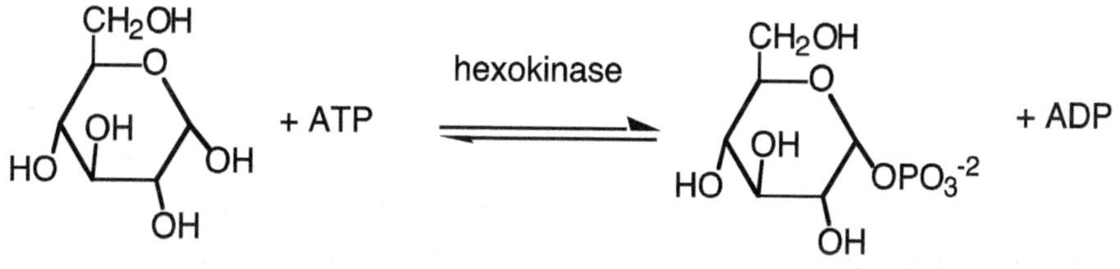

   hexokinase

   + ATP ⇌ + ADP

   b. Glucosamine is an inhibitor of hexokinase.
   What type of inhibitor would you expect it to be? (*Circle one.*)

   competitive noncompetitive

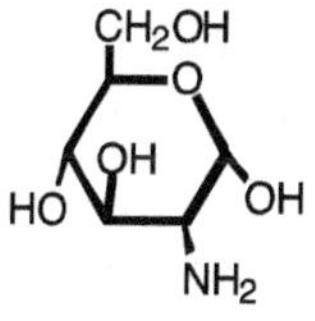

   Glucosamine

3. Jell-O® gelatin packages have the caution, "Do not use fresh or frozen pineapple, ... Gelatin will not set." Pineapple contains the enzyme bromelain that hydrolyzes the proteins in gelatin, destroying the gelling action. Canned pineapple can easily be used in gelatin. Why?

4. Lipid storage diseases occur in people who are deficient in enzymes to hydrolyze various sphingolipids.

Accumulating Lipid	Missing Enzyme	Disease Name
Galactocerebroside	β-Galactosidase	Krabbe's leukodystrophy
Ceramide trihexoside	α-Galactosidase	Fabry's disease
Sphingomyelin	Sphingomyelinase	Niemann-Pick disease
Glucocerebroside	β-Glucosidase	Gaucher's disease

Source: F. A. Bettelheim, W. H. Brown, M. K. Campbell, S. O. Farrell, *Introduction to General, Organic and Biochemistry*, 8th Ed., Saunders College Publishing, 2007, p. 534–5.

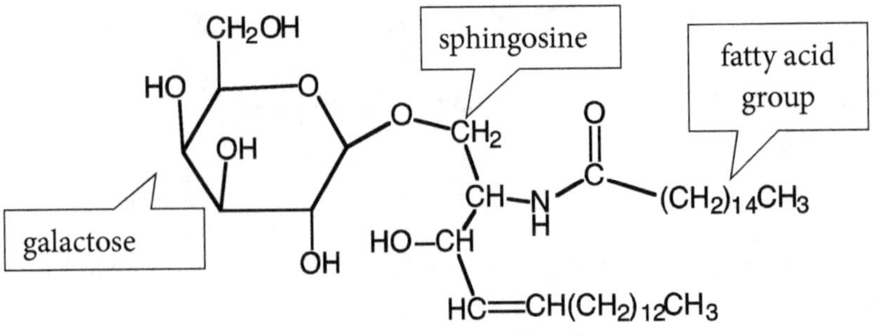

a Galactocerebroside

a. *Circle* the linkage broken by the enzyme missing in Krabbe's leukodystrophy.

b. To which enzyme class would this enzyme belong?

5. What molecule might be a competitive inhibitor for the enzyme missing in:
   a. Fabry's disease?

   b. Gaucher's disease?

6. Some contact lens cleaning solutions are described as enzymatic cleaners because they can remove protein deposits from lenses. One preparation contains proteases in combination with Endoproteinase lys-C, which are effective in dissolving away and hydrolyzing lysozyme, the major protein component of tears.

   To what class of enzymes would Endoproteinase lys-C belong?

7. Be sure you *understand* the answers to the ***Critical Thinking Questions*** and ***Exercises*** in this activity. *Ask more questions* until you are confident in your answers.

8. Read the corresponding sections and work the suggested problems in the text.

14

NUCLEIC ACIDS

How are organic reactions used in nucleic acid polymers?

Learning Objectives

- Identify the characteristic components in nucleic acids.
- Recognize the linkages in nucleic acids.
- Describe primary structures of DNA and RNA.

Prerequisite Concepts

- Recognition of esters, acetals, phosphate esters, phosphoric anhydrides
- Acid-base recognition
- Hydrolysis of specific functional groups
- Monomers, polymers, repeating units
- Numbering in carbohydrates

Model 1. Nucleosides

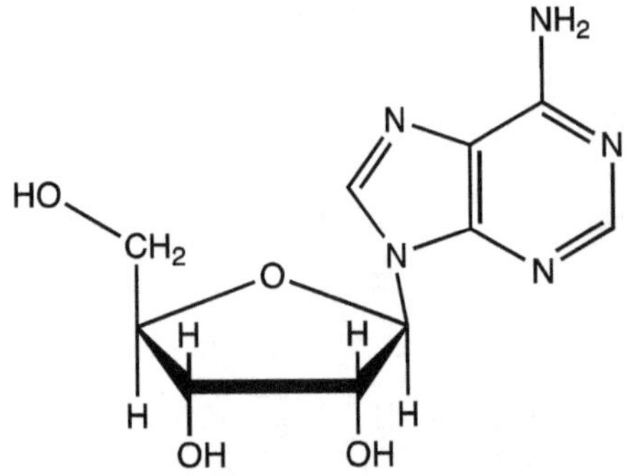

a) a ribonucleoside: ribose + adenine

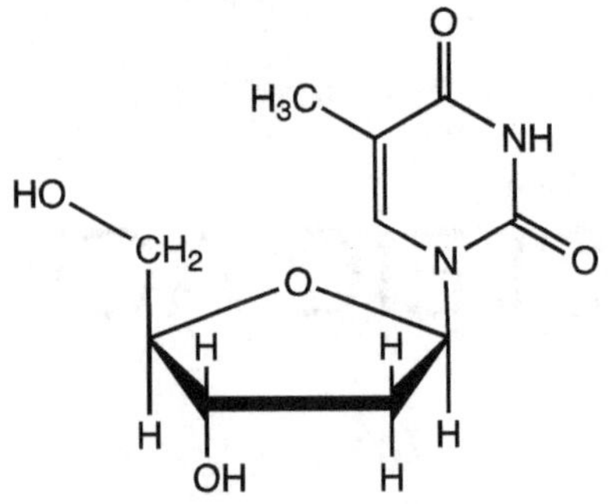

b) a deoxyribonucleoside: deoxyribose + thymine

Critical Thinking Questions

1. Number the carbons of the sugar in each structure of Model 1. Which carbon is connected to the nitrogen heterocycle?
   In a. In b.

2. Identify the sugars present.
   In a. In b.

3. What type of linkage connects the sugar to the nitrogen heterocycle? (*Circle one.*)

   amide N-acetal ester phosphate ester phosphoric anhydride

4. *As a group,* determine why the compounds in Model 1 might be described as nucleosides. (*Hint:* Think about other compounds e.g., in Activity 11, *Reactions of Carbohydrates*, described as *–osides*).

5. Adenine and thymine are described as bases in nucleic acids. *As a group,* decide what characteristic qualifies them as bases.

6. In general, what two components are present in a nucleoside? ___________ ___________

Information, Part I: Numbering in Nucleic Acids

Sugars in nucleic acids are given "primed" numbers, 1′, 2′, etc., to distinguish their numbering from positions on the bases.

7. Re-label the numbers on the Model 1 sugars with primed numbers.

Model 2. Nucleotides

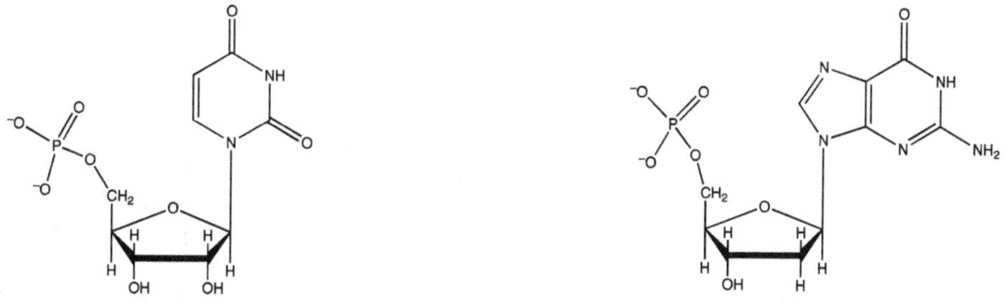

a) a ribonucleotide =
ribonucleoside 5'-monophosphate (NMP)
e.g. ribose + uracil + phosphate

b) a deoxyribonucleotide =
deoxyribonucleoside 5'-monophosphate (dNMP)
e.g. deoxyribose + guanine + phosphate

Critical Thinking Questions (cont'd.)

8. a. Number the carbons of each sugar in Model 2 using "primed" numbers. Which carbon is connected to the nitrogen heterocycle?

   In a. In b.

   b. Is the sugar-base linkage different from that in the sugars of Model 1? (*Circle one.*) Yes/No

9. Which carbon is connected to the phosphate group? In a. ______ In b. ______

10. What type of linkage connects the sugar to the phosphate?

    N-acetal amide ester phosphate ester phosphoric anhydride

11. What are the three components of a nucleotide?

    1. 2. 3.

12. *As a group*, determine how a ribonucleotide differs from a deoxyribonucleotide. (*Hint:* It's not the bases.)

13. a. If the nucleotide in Model 2a is known as UMP, what would it be called if the base is replaced by adenine?

    b. If the nucleotide in Model 2b is known as dGMP, what would it be called if the base is replaced by thymine?

Model 3. Nucleoside Triphosphates

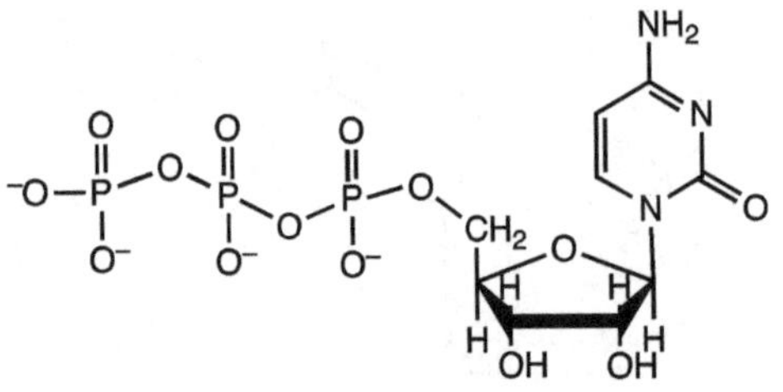

a) a ribonucleoside triphosphate (NTP)
ribose + cytosine + 3 phosphates

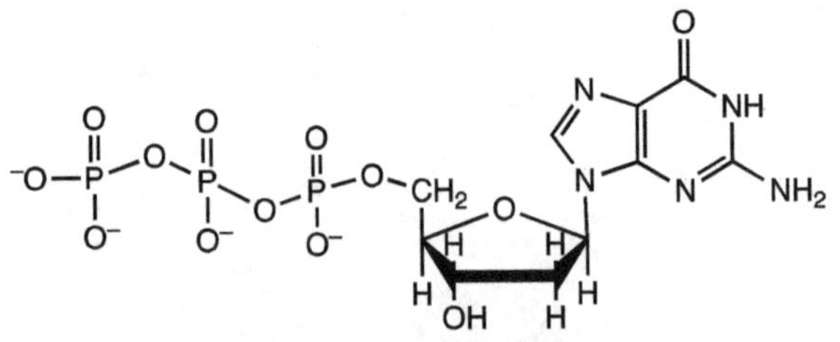

b) a deoxyribonucleoside triphosphate (dNTP)
deoxyribose + guanine + 3 phosphates

CRITICAL THINKING QUESTIONS (CONT'D.)

14. In a nucleoside triphosphate (NTP), what type of linkage is found between the phosphate groups? (*Circle one.*)

N-acetal amide ester phosphate ester phosphoric anhydride

15. a. If the compound in Model 3a is known as CTP, what would it be called if the base is replaced by adenine?

b. If the compound in Model 3b is known as dGTP, what would it be called if the base is replaced by thymine?

16. *As a group,* determine the difference between a nucleoside triphosphate and a nucleotide.

17. Identify the bases present:
Model 1a: ____________ Model 1b: ____________

Model 2a: ____________ Model 2b: ____________ Model 3a: ____________

Model 4. Nucleic Acid Polymers

a) a ribonucleotide polymer fragment

b) a deoxyribonucleotide polymer fragment

CRITICAL THINKING QUESTIONS (CONT'D.)

18. *As a group,* describe the repeating pattern in the backbone of the nucleic acid fragments of Model 4 in terms of nucleotide components. (*Hint:* See CTQ 11.)

19. *Number* the carbons (i.e., 1′, 2′, etc.) on the top and bottom sugars in each polymer fragment. One end of the fragment is known as the *5′ end* and the other is the *3′ end. Label* the ends of each fragment accordingly.

20. a. What group is attached at the 5′ end of each fragment?

    b. What group is attached at the 3′ end of each fragment?

21. Which might describe the linkages on the phosphates linked between two sugars in each fragment? (*Circle one.*)

    N- acetal ester phosphate ester phosphate diester phosphoric anhydride

22. *As a group,* explain why the nucleotide is described as the *repeating unit* of each nucleic acid polymer.

23. If the NTP molecules are the *monomers* from which a nucleic acid polymer is made, what by-products must result when the polymer is assembled?

Information Part II: Primary Structure in Nucleic Acids

The bases, the parts that make each unique, characterize each segment of DNA or RNA. Just as the side chains of amino acids distinguish them from one another, the bases of nucleic acids are the distinguishing characteristic. For proteins, the *sequence* of the amino acids (distinguished by the side chains) in the polypeptide describes the primary structure. Similarly, for nucleic acids the sequence of the bases (ACGT, etc.) in the polynucleotide describes the primary structure.

Critical Thinking Question (cont'd.)

24. Starting with the 5' end, and using information from Models 1–3, deduce the base sequence of the trinucleotides of Model 4.

    4a. (5′)____ ____ ____(3′) 4b. (5′) ____ ____ ____(3′)

Model 5. Aminoacyl End of $tRNA^{Phe}$

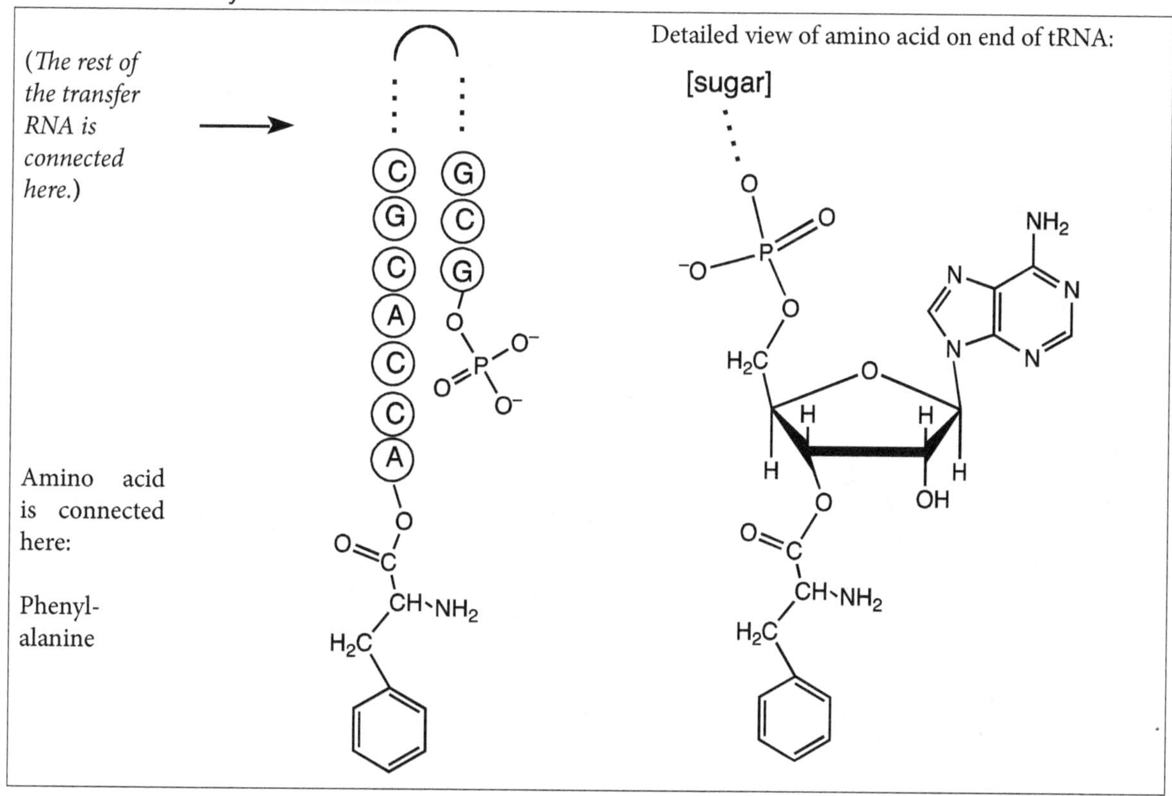

Critical Thinking Questions (cont'd.)

25. a. Which end of an RNA has a phosphate group? (*Circle one.*) (*Hint:* See CTQ 20.)

    5′ 3′

    b. Which end of the RNA must be attached to the amino acid? (*Circle one.*)

    5′ 3′

    c. *Discuss and decide as a group:* What is otherwise found on the end of a nucleic acid polymer as noted in CTQ 25b? (*Hint:* See *CTQ 20.*)

    d. Which functional group of the amino acid must be attached to the RNA? (*Circle one.*)

    α carbon amino group carboxylic acid group side chain

    e. Is this a group common to all amino acids? (*Circle one.*) Yes/No

    f. What type of linkage connects the amino acid to the end of the RNA? (*Circle one.*)

    amide N-acetal ester phosphate ester phosphoric anhydride

26. *As a group,* determine which linkages present in nucleic acids are subject to hydrolysis. *Circle* those that are hydrolyzable. (*Refer to CTQ 3, 9, 14, and 21 for linkages.*)

    base–sugar   sugar–phosphate   phosphate–phosphate   sugar–amino acid

Exercises

1. In the compound below:
   a. *Circle* the N-acetal linkage, and
   b. *Draw* a box around the phosphate ester.

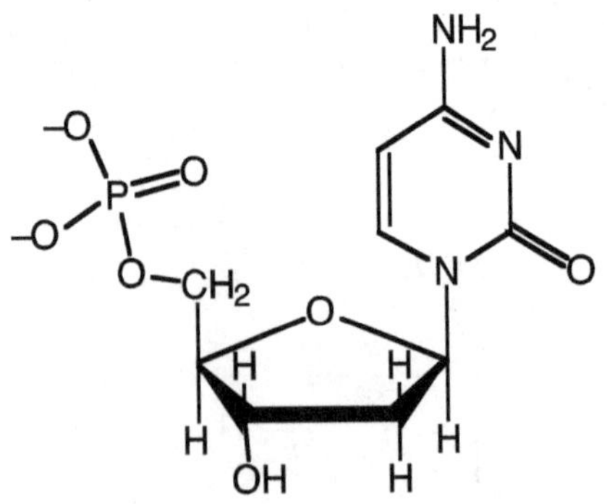

   c. Name the sugar present.

   d. To which carbon number is the phosphate attached?

   e. What group is attached at the 3′ carbon?

2. Comparing to structures in Models 1–4, which *base* is present in the compound of Exercise 1?

3. Which term applies to the compound in Exercise 1? (*Circle one.*)

   ribonucleoside   ribonucleotide   deoxyribonucleoside   deoxyribonucleotide

4. What would be the products of complete hydrolysis of these compounds?

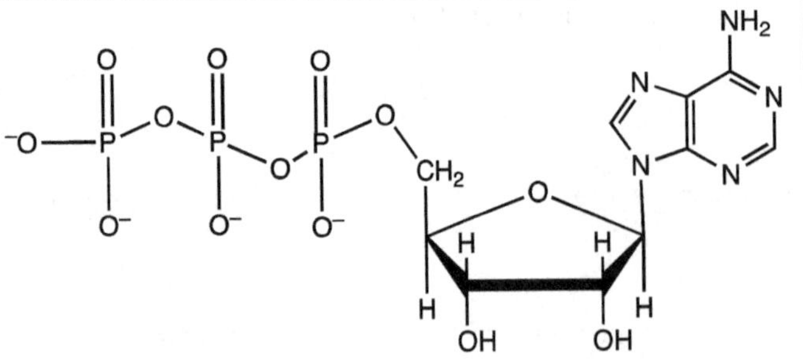

a) hydrolysis products:	b) hydrolysis products:

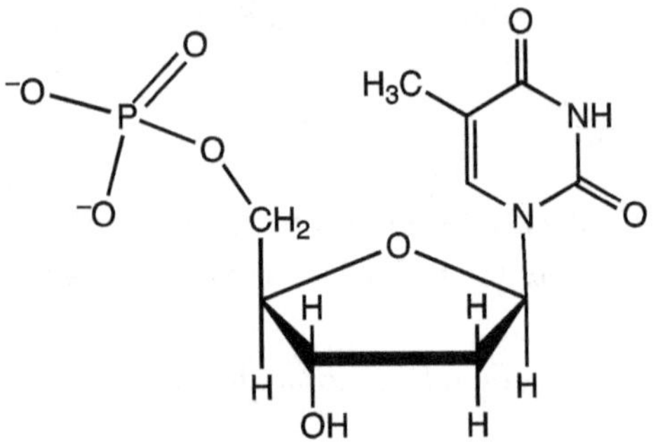

5. Two *bases* in nucleic acids differ only by one methyl ($-CH_3$) group.
   a. Which two are they?

   b. Which of these two includes the distinguishing methyl group?

6. Some bases in nucleic acids have double ring nitrogen heterocycles known as *purines*. Which bases are these?

7. Some bases in nucleic acids have single ring nitrogen heterocycles known as *pyrimidines*. Which bases are these?

8. *Circle* the correct choices for each box in the table.

Polymer	Type	Composed	of	Backbone connected by which linkage type?
RNA	linear or branched	deoxy- or ribo-	nucleosides or nucleotides	N-acetal amide ester phosphodiester phosphoric anhydride
DNA	linear or branched	deoxy- or ribo-	nucleosides or nucleotides	N-acetal amide ester phosphodiester phosphoric anhydride

9. How is the primary structure described for a segment of RNA or DNA? What types of linkages hold the structures together?

10. Be sure you *understand* the answers to the ***Critical Thinking Questions*** and ***Exercises*** in this activity. *Ask more questions* until you are confident in your answers.

11. Read the corresponding sections and work the suggested problems in the text.

15

THE CITRIC ACID CYCLE

What features of this oxidative cycle generate energy?

Learning Objectives:

- Identify important features of Coenzyme A.
- Explain the roles of coenzymes in the Citric Acid Cycle.
- Recognize reaction types at specific steps in the Citric Acid Cycle.

Prerequisite Concepts:

- Oxidation-reduction reactions
- Monosaccharide identification
- Enzyme classes, coenzymes, vitamins
- Coupled reactions and their representations
- Recognition of organic linkages, nucleotides
- ATP, adenosine triphosphate, an energy carrier

Model 1. Coenzyme A, a Vitamin-derived Molecule

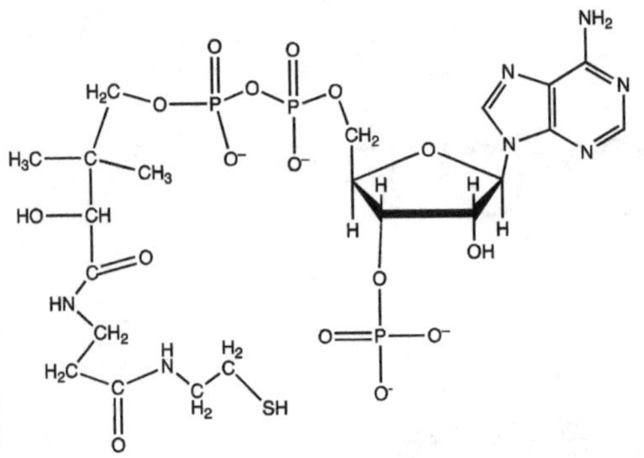

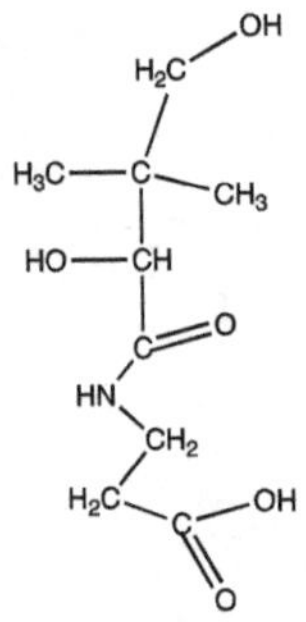

pantothenic acid, a vitamin

CRITICAL THINKING QUESTIONS

1. Note the parts comprising this coenzyme: a vitamin-derived portion, three phosphates, a sugar, a base, and 2-aminoethanethiol.

   a. *Circle* the monosaccharide present in Model 1 and identify it.

   deoxyribose   fructose   galactose   glucose   ribose

   b. Make a *box* around the vitamin-derived portion. Draw an *arrow* pointing to the base.

2. Name another molecule that contains the same sugar as found in Coenzyme A (CoA).

3. What is the name of the nitrogenous heterocyclic base in CoA? (*Hint*: It gives the A to the coenzyme's name.)

4. Part of CoA can be described as a *nucleotide*. (*Note:* Refer to Activity 14, *Nucleic Acids*, for a definition.) Which parts does that include?

5. *As a group,* determine what type of linkage connects the pantothenic acid to the phosphate in this molecule. (*Circle one.*)

acetal   amide   ester   phosphate ester   thioester

Model 2. Acetyl Coenzyme A

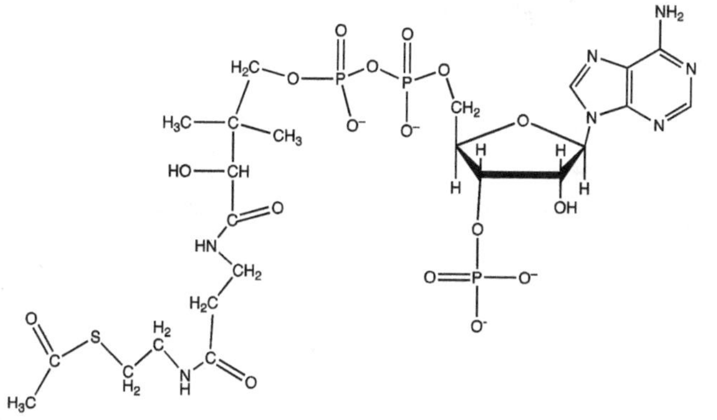

The reactive *thiol* group of CoA is sometimes explicitly represented as CoA-**SH**. This thiol can covalently carry an acetyl group ($CH_3C{=}O$) in a *thioester* bond, represented as acetyl-S-CoA or Ac-S-CoA:

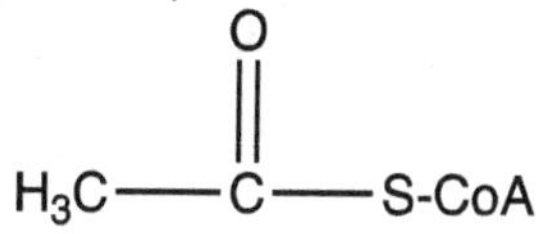

$$H_3C-\overset{\overset{\displaystyle O}{\|}}{C}-S\text{-}CoA$$

CRITICAL THINKING QUESTION (CONT'D.)

6. *Circle* the thioester linkage in Model 2. *As a group,* determine how it is similar to or different from a carboxylic ester.

Model 3. The Citric Acid Cycle

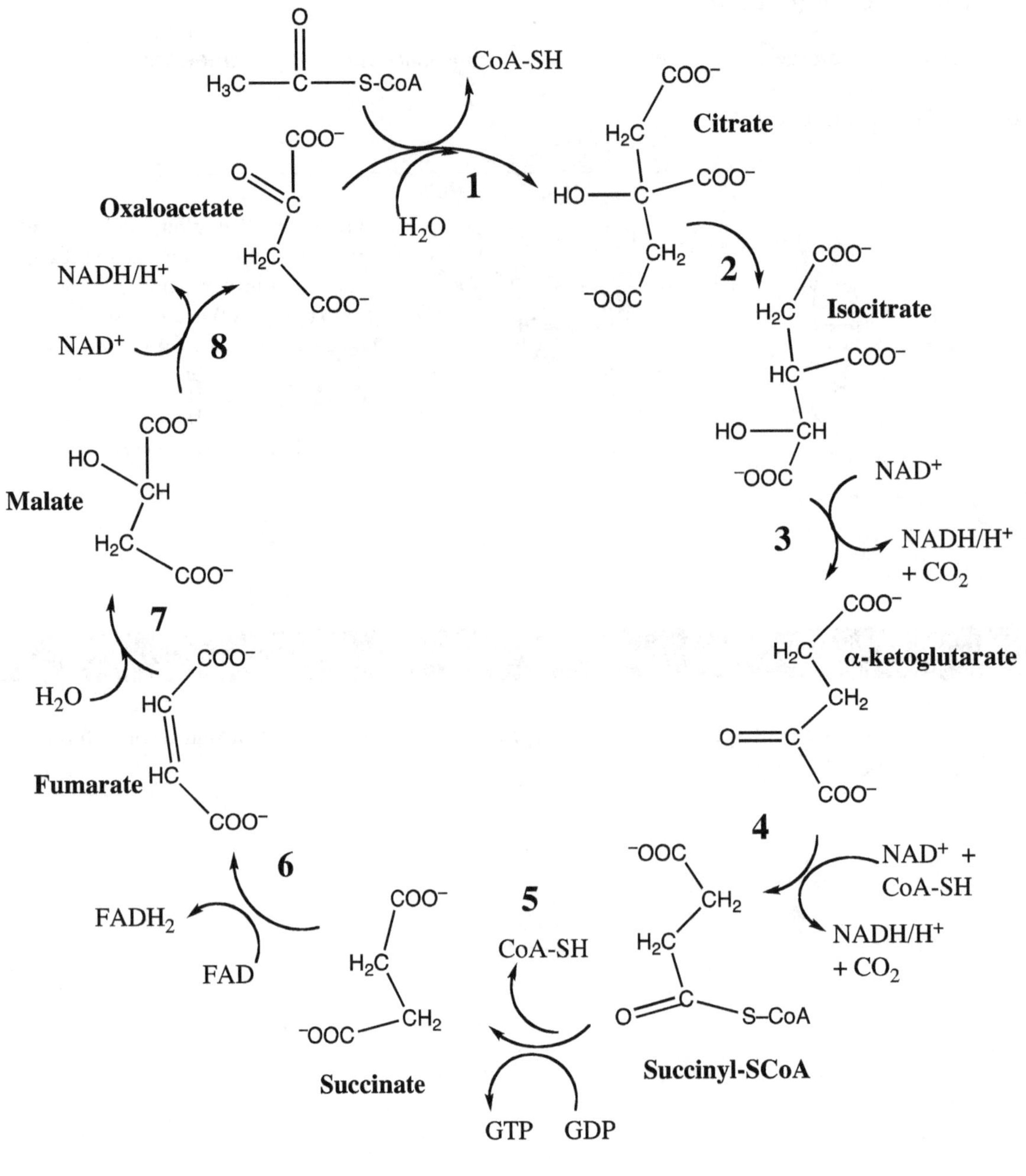

CRITICAL THINKING QUESTIONS (CONT'D.)

(*Note:* Refer to Model 3.)

7. Looking at the product of step 1, why might this cycle sometimes be referred to as the Tri-Carboxylic Acid Cycle?

8. The Citric Acid Cycle (CAC) requires the *oxidizing* coenzymes NAD^+ and FAD that become *reduced* to $NADH/H^+$ and $FADH_2$, respectively.

   a. Which are the numbers of the steps that involve NAD^+?

   b. Which are the number(s) of the steps that involve FAD?

9. *High-energy* molecules that result from the CAC include NADH and $FADH_2$. How many of each of these reduced coenzymes result from one turn (all 8 steps) of the cycle? (*Circle one for each coenzyme.*)

   a. Number of NADH produced per turn of the CAC: 1 2 3

   b. Number of $FADH_2$ produced per turn of the CAC: 1 2 3

10. Each of the eight steps of the CAC requires enzymes.

    a. Which steps would involve an *oxidoreductase* enzyme?

    (*There are 4.*) _____ _____ _____ _____

    b. *As a group,* decide what good clue indicates that oxidation occurs at these steps.

11. Step 2 of the cycle converts citrate to isocitrate.
    a. What term describes how these compounds are related?

    b. What is different about the alcohol groups in the two compounds? (*Hint:* How are they classified?)

    c. Which class of enzymes would likely be involved in Step 2? (*Circle one.*)
    hydrolase isomerase ligase lyase oxidoreductase transferase

12. Tertiary alcohols are not oxidizable. (*Hint:* Recall the oxidation of alcohols.) In the context of the series of reactions to follow in the CAC, *decide as a group* what the benefit of the reaction in Step 2 might be.

13. In which step do two carbons enter the cycle? _________

14. Each turn of the CAC results in the production of two CO_2 molecules. In which steps is CO_2 released?

15. In Step 1 of the CAC, an acetyl group is transferred from Acetyl-CoA to oxaloacetate. The product of Step 8 is oxaloacetate. Why is this metabolic pathway referred to as *cyclic*?

16. Step 7 is the addition of H_2O to an alkene resulting in the formation of an alcohol. Which class of enzymes would be involved at this step? (*Hint:* Refer to Model 3 of Activity 13, *Enzymes.*) *Confer with your group.* (*Circle one.*)

    hydrolase   isomerase   ligase   lyase   oxidoreductase   transferase

17. No ATP is produced *directly* from the CAC, but *one* similar molecule GTP (*guanosine* triphosphate) is formed in Step 5. The phosphorylation of one ADP to ATP is *coupled* with the hydrolysis of GTP (from Step 5) to GDP. Write the reaction for this coupled process using the curved arrow notation.

18. *As a group,* decide why the formation of ATP from Step 5 qualifies as *substrate-level phosphorylation.*

Exercises

1. Write a chemical equation for these steps of the CAC. (The first is done for you.) Names of compounds may be used in place of chemical structures.

   Step 1: Oxaloacetate + Acetyl-SCoA + H_2O $\longrightarrow$ citrate + CoA-SH

   a. Step 2:

   b. Step 4:

   c. Step 6:

   d. Step 7:

2. a. Which compound provides two carbons to the Citric Acid Cycle in Step 1?

   b. From what food source(s) does it originate? (*Recall from or find in your text.*)

3. List the following molecules as required for or products of one turn of the Citric Acid Cycle.

CoA-SH	Ac-S-CoA	3 NAD^+	3 $NADH/H^+$	FAD	$FADH_2$
ATP	ADP	HPO_4^{-2}	H_2O	2 CO_2	GDP GTP

Required Molecules:	Products:

4. The oxidations in the CAC that require NAD^+ or FAD are of two specific types. Which *coenzyme* is used for each of these?

   a) Oxidation of a 2° alcohol (by loss of 2 H) to make a ketone: ____________

   b) Oxidation of a single C–C (by loss of 2 H) to yield a double C=C bond: __________

5. Two steps of the CAC are described as *oxidative decarboxylation.*

   a. Which two steps are they?

   b. What qualifies these steps as oxidative?

   c. What qualifies these steps as decarboxylation?

6. The coenzyme NAD^+, nicotine adenine dinucleotide, is shown below.

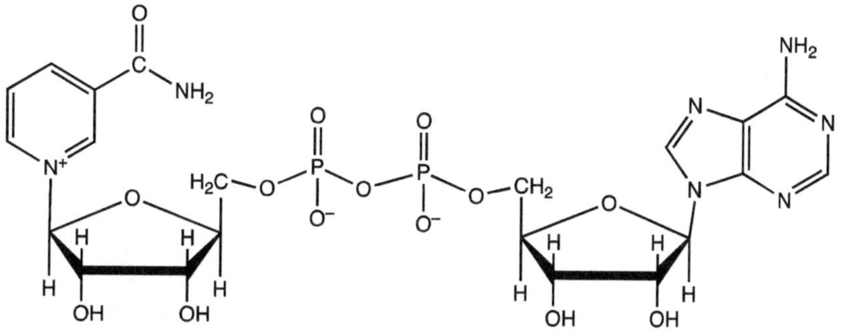

   a. *Circle* adenine in the structure.
   b. What sugar is present?
   c. Draw an *arrow* to the positive charge on NAD^+.
   d. Put an asterisk (*) by the nicotinamide group, the second N-heterocycle.
   e. What qualifies the structure as a *di*-nucleotide? Put a set of *brackets* [ ] around each nucleotide.

7. The coenzyme FAD, flavin adenine dinucleotide, is shown below.

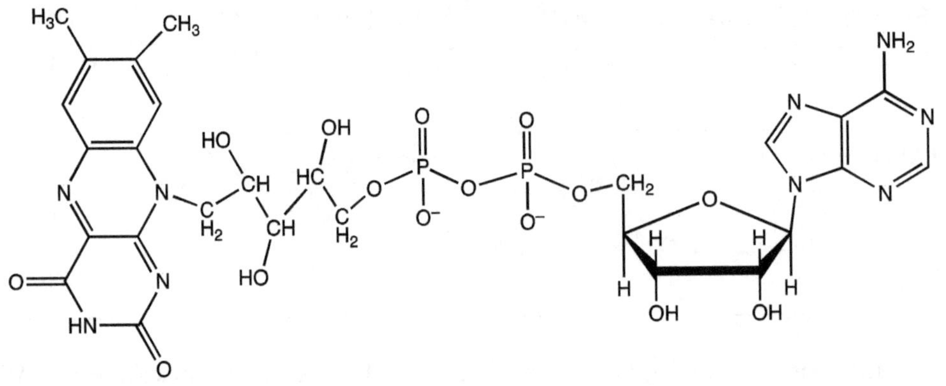

   a. *Circle* adenine in the structure. Put an asterisk (*) by flavin, the second N-heterocycle.
   b. What sugar is present?

   c. What qualifies the structure as a *di*-nucleotide? Put a set of *brackets* [ ] around each nucleotide.

8. From what nutritional source(s) are nicotinamide and flavin derived? (*Hint:* Look for vitamins in your text.)

9. Looking at the structures of Coenzyme A, NAD^+, and FAD, three coenzymes in the Citric Acid Cycle, what features are common to all three coenzymes?

10. Be sure you *understand* the answers to the ***Critical Thinking Questions*** and ***Exercises*** in this activity. *Ask more questions* until you are confident in your answers.

11. Read the corresponding sections and work the suggested problems in the text.

CPSIA information can be obtained
at www.ICGtesting.com
Printed in the USA
LVHW060141110321
681116LV00003B/10